Nucleosides, Nucleotides, and Their Biological Applications

Nucleosides, Nucleotides, and Their Biological Applications

EDITED BY

Janet L. Rideout
David W. Henry
Lowrie M. Beacham III

Burroughs Wellcome Co.
Research Triangle Park, North Carolina

Proceedings of the 5th International Round Table
October 20–22, 1982

1983

ACADEMIC PRESS
A Subsidiary of Harcourt Brace Jovanovich, Publishers
New York   London
Paris   San Diego   San Francisco   São Paulo   Sydney   Tokyo   Toronto

ACADEMIC PRESS, INC.
111 Fifth Avenue, New York, New York 10003

United Kingdom Edition published by
ACADEMIC PRESS, INC. (LONDON) LTD.
24/28 Oval Road, London NW1 7DX

Library of Congress Cataloging in Publication Data

Main entry under title:

Nucleosides, nucleotides, and their biological
    applications.

    Includes index.
    1. Nucleosides–Congresses. 2. Nucleotides–Con-
gresses. I. Rideout, Janet L. II. Henry, David W.
III. Beacham III, Lowrie M.
QP625.N88N83   1983   599'.087328   83-15745
ISBN 0-12-587980-6 (alk. paper)

PRINTED IN THE UNITED STATES OF AMERICA

83 84 85 86   9 8 7 6 5 4 3 2 1

Contents

Contributors

Numbers in parentheses indicate the pages on which the authors' contributions begin.

J. H. Burchenal (47), Memorial Sloan-Kettering Cancer Center, New York, New York

Robert F. Bruns (117), Department of Pharmacology, Warner-Lambert/Parke Davis Pharmaceutical Research, Ann Arbor, Michigan

Werner Bussman (181), Laboratory of Bioorganic Chemistry, Department of Chemistry, University of Paderborn, D-4790 Paderborn, Federal Republic of Germany

Ramamurthy Charubala (147), Fakultät Für Chemie, Universität Konstanz, Konstanz, Germany

Claude Chavis (257), Bio-Organic Laboratory, University of Sciences and Techniques of Languedoc, Montpellier, France

Thomas L. Cupps (297), Department of Medicinal Chemistry, University of Michigan, Ann Arbor, Michigan

Yair Devash (147), Department of Biochemistry, Temple University School of Medicine, Philadelphia, Pennsylvania

Paul Doetsch (147), Department of Biochemistry, Temple University School of Medicine, Philadelphia, Pennsylvania

Deborah A. Eppstein (257), Institute of Bio-Organic Chemistry, Syntex Research, Palo Alto, California

J. J. Fox (47), Memorial Sloan-Kettering Cancer Center, New York, New York

Doris Franzen (181), Laboratory of Bioorganic Chemistry, Department of Chemistry, University of Paderborn, D-4790 Paderborn, Federal Republic of Germany

Gilles Gosselin (257), Bio-Organic Laboratory, University of Sciences and Techniques of Languedoc, Montpellier, France

Abd El Fattah Haïkal (257), Bio-Organic Laboratory, University of Sciences and Techniques of Languedoc, Montpellier, France

Fritz Hansske (279), Department of Chemistry, The University of Alberta, Edmonton, Alberta, Canada

Doris Hasselmann (181), Laboratory of Bioorganic Chemistry, Department of Chemistry, University of Paderborn, D-4790 Paderborn, Federal Republic of Germany

S. D. Hawrelak (279), Department of Chemistry, The University of Alberta, Edmonton, Alberta, Canada

Earl E. Henderson (147), Department of Microbiology and Immunology, Fels Research Institute, Temple University School of Medicine, Philadelphia, Pennsylvania

Jiro Imai (67), Laboratory of Chemistry, National Institute of Arthritis, Diabetes, and Digestive and Kidney Diseases, National Institutes of Health, Bethesda, Maryland

Jean-Louis Imbach (257), Bio-Organic Laboratory, University of Sciences and Techniques of Languedoc, Montpellier, France

Margaret I. Johnston (67), Laboratory of Chemistry, National Institute of Arthritis, Diabetes, and Digestive and Kidney Diseases, National Institutes of Health, Bethesda, Maryland

R. Klein (47), Memorial Sloan-Kettering Cancer Center, New York, New York

John C. Krauss (297), Department of Medicinal Chemistry, University of Michigan, Ann Arbor, Michigan

Krystyna Lesiak (67), Laboratory of Chemistry, National Institute of Arthritis, Diabetes, and Digestive and Kidney Diseases, National Institutes of Health, Bethesda, Maryland

B. Leyland-Jones (47), Memorial Sloan-Kettering Cancer Center, New York, New York

C. Lopez (47), Memorial Sloan-Kettering Cancer Center, New York, New York

Danuta Madej (279), Department of Chemistry, The University of Alberta, Edmonton, Alberta, Canada

Y. Vivienne Marsh (257), Institute of Bio-Organic Chemistry, Syntex Research, Palo Alto, California

John A. Montgomery (19), Organic Chemistry Research Department, Southern Research Institute, Birmingham, Alabama

Kelvin K. Ogilvie (209), Department of Chemistry, McGill University, Montreal, Quebec, Canada

Johann Ott (181), Laboratory of Bioorganic Chemistry, Department of Chemistry, University of Paderborn, D-4790 Paderborn, Federal Republic of Germany

Wolfgang Pfleiderer (147), Fakultät Für Chemie, Universität Konstanz, Konstanz, Germany

Morris J. Robins (279), Department of Chemistry, The University of Alberta, Edmonton, Alberta, Canada

Hiroaki Sawai (67), Faculty of Pharmaceutical Sciences, University of Tokyo, Bunkyo-ku, Tokyo, Japan

Howard J. Schaeffer (1), Department of Development, Burroughs Wellcome Co., Research Triangle Park, North Carolina

Brian B. Schryver (257), Institute of Bio-Organic Chemistry, Syntex Research, Palo Alto, California

Frank Seela (181), Laboratory of Bioorganic Chemistry, Department of Chemistry, University of Paderborn, D-4790 Paderborn, Federal Republic of Germany

Robert J. Suhadolnik (147), Department of Biochemistry, Temple University School of Medicine, Philadelphia, Pennsylvania

Paul F. Torrence (67), Laboratory of Chemistry, National Institute of Arthritis, Diabetes, and Digestive and Kidney Diseases, National Institutes of Health, Bethesda, Maryland

Leroy B. Townsend (297), Department of Medicinal Chemistry, University of Michigan, Ann Arbor, Michigan 48109

Quynh-Hoa Tran-Thi (181), Laboratory of Bioorganic Chemistry, Department of Chemistry, University of Paderborn, D-4790 Paderborn, Federal Republic of Germany

B. Watanabe (47), Memorial Sloan-Kettering Cancer Center, New York, New York

John S. Wilson (279), Department of Chemistry, The University of Alberta, Edmonton, Alberta, Canada

Heinz-Dieter Winkeler (181), Laboratory of Bioorganic Chemistry, Department of Chemistry, University of Paderborn, D-4790 Paderborn, Federal Republic of Germany

Dean S. Wise (297), Department of Medicinal Chemistry, University of Michigan, Ann Arbor, Michigan 48109

Joseph M. Wu (147), Department of Biochemistry, New York Medical College, Valhalla, New York

Preface

The Fifth International Round Table on Nucleosides, Nucleotides, and Their Biological Applications was held at Research Triangle Park, North Carolina on October 20–22, 1982. The program consisted of 11 plenary lectures interspersed with poster presentations and attracted approximately 125 attendees. This volume presents the manuscripts on which the distinguished group of plenary lecturers based their presentations.

A word on the sponsorship of this series of Round Table meetings might be of interest. This fifth meeting was sponsored by Burroughs Wellcome Co., in whose headquarters the meeting took place. Previous sponsors were the University of Antwerp (R. U. C. A.) (Antwerp, Belgium, 1981), the U. S. T. L. (Montpellier, France, 1978), The American Chemical Society (San Francisco, 1976), and again the U. S. T. L. in Montpellier (1974) where the initial Round Table took place. This meeting is unique in being focused on a small group of scientists united by a common interest in nucleosides and nucleotides but without a specific organizational affiliation. Especially important in initiating and encouraging the continuance of this series have been Professors J. L. Imbach of Montpellier and Leroy B. Townsend of The University of Michigan.

The Round Tables provide a forum at which chemists and biologists with a passion for this specialty may present new data, espouse new hypotheses, air differences, and generally maintain the rapport that has resulted in the nucleoside research field being the exciting scientific arena that it is today. The subject matter of the Fifth Round Table spanned a number of subspecialties and areas of unusually intense current interest. Continuing advances in synthetic chemistry were well represented as were fast-moving biochemical subjects such as the 2',5'-oligo-adenylates and their relationship to interferons. Rapidly expanding interest in adenosine and adenine nucleotide receptors, and in nucleosides with selective antiviral action, also stimulated important elements in the program. The quality of the speakers and their presentations was uniformly outstanding, with exceptionally vigorous participation by attendees in discussions both following presentations and during the breaks.

The editors especially wish to thank Mrs. Blenda G. Morris, of Burroughs Wellcome Co., who served ably as Round Table secretary; Ms. Lilia M. Beauchamp and Dr. James L. Kelley, who arranged the poster sessions and assembled the abstracts; Dr. Tom R. Henderson, the tour chairman; and Ms. Jean F. Frazier, who handled so many of the arrangements as administrative assistant and who served so well as technical editor for this volume of proceedings. We also wish to express our gratitude to Burroughs Wellcome Co. for the use of its meeting organization services and facilities and for financial support. Most of all we thank the speakers for their hard work and for excellent science, the keys to any successful meeting.

NUCLEOSIDES WITH ANTIVIRAL ACTIVITY[1]

Howard J. Schaeffer

Department of Development

Burroughs Wellcome Co.

Research Triangle Park, NC

Through 1981, there were four antiviral agents approved
for use in the United States by the FDA. Amantadine was
approved as a prophylactic agent and three nucleosides
were approved as curative agents. The nucleoside antiviral
agents are idoxuridine (5-iodo-2'-deoxyuridine, IUDR),
vidarabine (adenine arabinoside, ara-A), and trifluridine
(5-trifluoromethyl-2'-deoxyuridine, TFT). In order for
these nucleosides to exert their antiviral effects, they
must be activated after administration to their 5-monophos-
phate form and ultimately to their triphosphate form. The
triphosphate form is an inhibitor of DNA-polymerase and is
also incorporated in DNA. Unfortunately, there is little
or no selectivity of activation so the antiviral agents
are incorporated into the DNA of the virus as well as into
the DNA of uninfected cells which, in turn, limits their
utility for systemic administration.

In recent studies, several new nucleoside analogs have
been discovered which show a high degree of potency and
selectivity against viruses of the herpes group. Three of
these potent antiviral agents are E-5-(2-bromovinyl)-2'-
deoxyuridine (BVDU), 1-(2-fluoro-2-deoxy-β-D-arabinofuran-
osyl)-5-iodocytosine (FIAC), and acyclovir (9-(2-hydroxy-

---

[1]No attempt has been made to review exhaustively the literature,
but general references are provided wherein additional references
can be found.

Nucleosides, Nucleotides,
and Their Biological Applications

1

ethoxymethyl)guanine, ACV. Acyclovir has recently been
approved by the FDA for topical and intravenous use.)
These newer compounds must also be activated to their
phosphate forms to exert their antiviral activities.
However, they are activated by specific enzymes coded by
the herpes virus. Thus, activation of these new agents
and inhibition of herpes replication occurs where it is
desired, i.e., only in infected cells. Details of the
mechanism of action are presented.

Selective toxicity is the key to the development of useful
chemotherapeutic agents. While there now exist many useful
antibacterial agents, there are relatively few clinically
effective antiviral agents. This disparity may be due to the
greater number of chemotherapeutic targets present in bacteria
compared to viruses, or it may reflect the disparity of research
effort that has been devoted to the development of antibacterial
agents compared to antiviral agents.

Prior to 1982, there were only four antiviral agents approved by the FDA for use in the United States. Except for amantadine (1-aminoadamantane hydrochloride) (1), the other three
antiviral agents are nucleoside analogs. These three antiviral
agents, which are used topically for the treatment of herpes
infections of the eye, are 5-iodo-2'-deoxyuridine(idoxuridine,
IUDR) (2), 9-β-D-arabinofuranosyladenine(vidarabine, ara-A),
and trifluorothymidine (TFT) (3). In addition, vidarabine has
recently been approved for use in the systemic treatment of
herpes encephalitis. During 1982, a fourth nucleoside analog,
9-(2-hydroxyethoxymethyl)guanine (acyclovir, ACV) (4,5) was
approved in the U.S. for topical and intravenous use in certain
herpes infections.

Because a number of nucleoside analogs have shown good
activity as anticancer and/or antiviral agents, there is no
doubt that continued research in nucleoside chemistry will be a
fruitful area of investigation in the search for compounds with
selective toxicity. In the antiviral field, a number of new
agents have been found with selective antiviral activity. One

TABLE I. Herpes Infections in Humans

---

1) HSV Type 1
    herpes labialis
    herpes keratitis
    herpes encephalitis

2) HSV Type 2
    herpes genitalis
    cervical neoplasia (?)

3) Varicella Zoster (VZV)
    chicken pox
    shingles

4) Cytomegalovirus (CMV)
    neonatal disease

5) Epstein-Barr Virus (EB)
    infectious mononucleosis
    Burkitt's lymphoma

---

group of viruses - the herpes group - appears to be particularly sensitive to a variety of nucleoside analogs. Consequently, this presentation will concentrate on nucleoside analogs with activity against the herpes group of viruses.

The herpes viruses that infect humans are comprised of five different groups. The viruses along with some of the infections that they cause in humans are outlined in Table I.

IUDR, ara-A, and TFT may be referred to as first-generation antiviral agents because these agents, while quite potent, do not exhibit a high degree of selective activity. Consequently, their use is limited to topical application except for the use

of ara-A in potentially lethal infections. These three compounds
are the forerunners of the newer antiviral agents, and they
exert their antiviral effect by a mechanism of action that is
apparently common to all the nucleoside antivirals. For example,
ara-A must be activated by several enzymes proceeding through
the 5'-monophosphate to the 5'-triphosphate (see Fig. 1). The
triphosphate of ara-A exerts a greater inhibitory effect on
HSV-DNA polymerase than it does on cellular DNA polymerase. In
addition, it is incorporated into DNA of virally-infected as
well as uninfected cells. The incorporation of ara-A into
cellular DNA may account for some of its toxicity. Furthermore,
ara-A suffers from the disadvantage that it is a good substrate
of adenosine deaminase such that systemically administered
ara-A is rapidly metabolized to the less active hypoxanthine
analog (6).

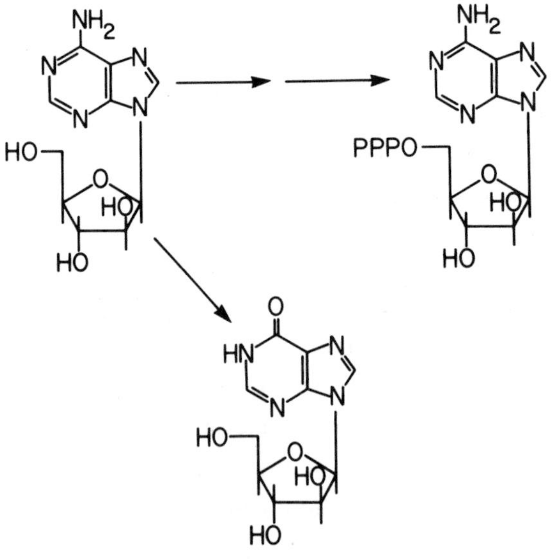

FIGURE 1.   Activation and metabolism of ara-A.   The letters
PPP represent the triphosphate moiety.

IUDR and TFT must also be activated to their triphosphate
forms to exert their antiviral effect (7,8). These compounds
also suffer from the disadvantage that they are activated both
in herpes infected and in uninfected cells. Their subsequent
incorporation into the DNA of uninfected cells limits their
utility for systemic administration. In the case of systemic
administration of TFT to animals, it undergoes phosphorolysis
to 5-trifluoromethyluracil and hydrolysis to 5-carboxy-2'-deoxy-
uridine, neither of which has significant antiviral activity.

Over the past several years, a number of highly selective
nucleoside antiviral agents have been prepared. Some of the
second generation of antiviral drugs include 5'-amino-5-iodo-
2',5'-dideoxyuridine (AIU) (9,10), 1-β-D-ribofuranosyl-1,2,4-
triazole-3-carboxamide (ribavirin) (11), 1-β-D-arabinofuranosyl-
thymine (ara-T) (12), several 5-alkyl substituted-2'-deoxyuri-
dines including E-5-(2-bromovinyl-2'-deoxyuridine (BVDU) (13),
1-(2-deoxy-2-fluoro-β-D-arabinofuranosyl)-5-iodocytosine (FIAC)
(14,15), 1-(2-deoxy-2-fluoro-β-D-arabinofuranosyl)-5-methyluracil
(FMAU) (14,15), and 9-(2-hydroxyethoxymethyl)guanine (acyclovir,
ACV) (4,5).

Before discussing selected members of these newer antiviral
agents, it is necessary to present a brief description of
several important events that occur in mammalian cells shortly
after they are infected by herpes virus, especially HSV-1,
HSV-2 and varicella zoster (16). Although the herpes virus
utilizes for its replication many of the natural enzymes present
in the host cell, it does induce the formation of enzymes that
are coded by the viral genome. Two of these enzymes, herpes-
encoded thymidine kinase (actually a deoxynucleoside kinase)
and herpes-encoded DNA-polymerase, play a key role in the
selective antiviral activity of the newer antiviral nucleoside
analogs. One of the first nucleoside antiviral agents to
exhibit truly selective activity against the herpes virus was

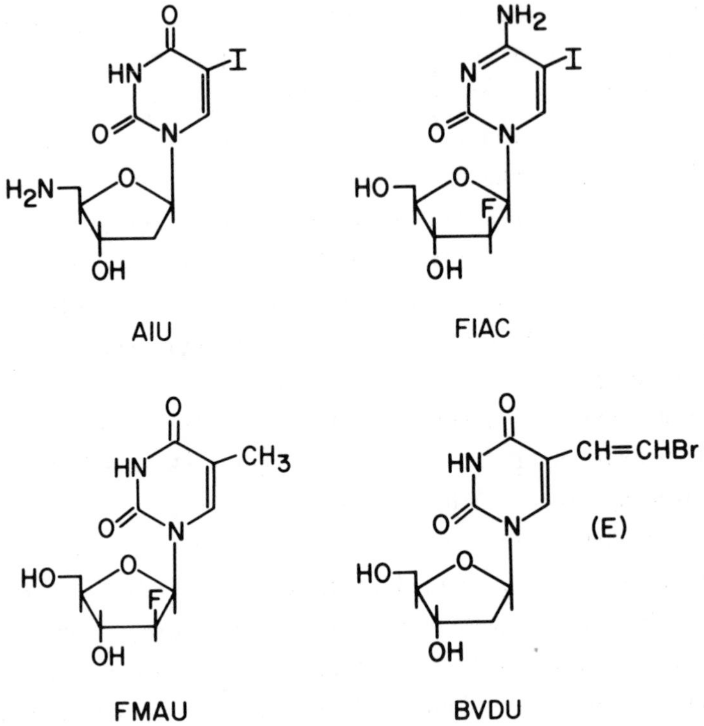

Howard J. Schaeffer

FIGURE 2.   Chemical structures of AIU, FIAC FMAU and BVDU.

AIU (Fig. 2). Prusoff and his coworkers (9,10,17) made the
exciting discovery that AIU is activated by the herpes-specified
thymidine kinase (TK), but AIU is not a good substrate of the
mammalian thymidine kinase. This compound serves as the first
example of a nucleoside antiviral agent with truly selective
activation. In HSV-1 infected cells, AIU is converted into its
triphosphoramidate form which is an inhibitor of DNA polymerase
and which is incorporated into the DNA. In uninfected cells,
however, little or no activation of AIU occurs; consequently,
the compound exhibits low toxicity to uninfected cells. Unfortunately, AIU is not very potent as an antiviral agent. Other
analogs of AIU are being prepared in an attempt to discover one
with increased potency for use in clinical trials.

Of a series of 2'-fluoro-2'-deoxyarabinofuranosylpyrimidine nucleosides, FIAC and FMAU (Figure 2) have been shown to be very potent and highly selective antiviral agents against HSV-1 and HSV-2 (14,15). Both FIAC and FMAU exhibit low cytotoxicity to uninfected cells grown in culture. Both of these compounds apparently exert their selective antiviral effects by being activated by a virus-specified thymidine kinase. Thus, when FIAC was studied with an HSV-1 mutant that lacked the virus-specified TK, it was found that the antiviral activity of FIAC was greatly reduced. Compared with the wild-type strain of HSV-1 (TK^{+}), the TK^{-} strain was approximately 8000 fold less sensitive to FIAC (18). Recent studies on isolated enzymes have shown that FIAC and FMAU serve as substrates for thymidine kinases isolated from HSV-1, HSV-2, and from human cells (19). In general, FIAC and FMAU were converted to their phosphate analogs at a greater rate by the virally-induced TK's than by the cytosol TK although FMAU exhibited a high degree of reaction with the cytosol TK. The importance of the latter finding is not yet established, but it is clear from in vivo studies that FIAC and FMAU are potent and selective anti-herpesvirus agents.

BVDU (Fig. 2) is another second generation antiviral agent with potent activity against some of the viruses of the herpes group (13,20).

BVDU has high activity against HSV-1 and VZV, but is markedly less inhibitory to HSV-2 (21). In common with the other nucleoside analogs with activity against the herpesviruses, BVDU is activated to its monophosphate by a virally-induced thymidine kinase. Subsequently, the monophosphate is converted into the triphosphate of BVDU which is a good inhibitor of HSV-1 DNA polymerase. Similarly, BVDU exerts its selective antiviral activity by virtue of the fact that it is a poor substrate for mammalian thymidine kinase.

Howard J. Schaeffer

9-(2-Hydroxyethoxymethyl)guanine (acyclovir, ACV) has been
shown to exhibit high activity against viruses of the herpes
group, especially against HSV-1, HSV-2, and varicella zoster
virus (4,5). Acyclovir may be regarded as a nucleoside analog
of guanosine in which the 2- and 3-carbon atoms of the cyclic
sugar moiety are missing. Earlier studies, using adenosine
deaminase as a model enzyme, had shown that the cyclic carbohydrate moiety of adenosine was not necessary to mimic nucleoside
binding to the enzyme (22). Based on this observation, a
series of acyclic nucleoside analogs were prepared with the
hope that they might have anticancer or antiviral activity.
From these studies, acyclovir emerged as one of the compounds
with potent antiviral activity (4,5).

Our initial synthesis of acyclovir involved the condensation
of 2,6-dichloropurine with 2-benzoyloxyethyl chloromethyl ether
which gave as the major product 2,6-dichloro-9-(2-benzoyloxy-
ethoxymethyl)purine (Fig. 3). Conversion of the 6-chloro group

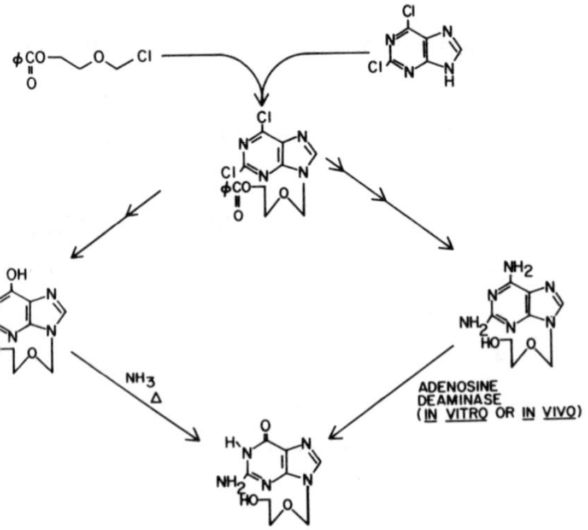

FIGURE 3. Synthesis of acyclovir (From Ref. 4).

to the oxygen function followed by ammonolysis at the 2-position
gave the desired acyclovir (4). Alternatively, 2,6-dichloro-9-
(2-benzoyloxyethoxymethyl)purine was converted via several
steps to the corresponding 2,6-diamino-9-(2-hydroxyethoxymethyl)-
purine which upon treatment with adenosine deaminase gave a
good yield of acyclovir. Subsequently, we and others have
devised more efficient procedures for the synthesis of acyclovir.
For example, Barrio, Bryant, and Keyser (23) have described an
elegant method for the synthesis of acyclovir (Fig. 4). Their
synthesis involved the reaction of 1,3-dioxolane with trimethylsilyl iodide which gave the desired blocked side-chain precursor.
This protected iodomethyl ether upon reaction with 2-chloro-6-
iodopurine gave 9-(2-hydroxyethoxymethyl)-2-chloro-6-iodopurine
which upon hydrolysis of the 6-iodo group followed by ammonolysis
of the 2-chloro group gave acyclovir.

Robins and Hatfield (24) have also devised an efficient
method for the synthesis of acyclovir (see Fig. 5). In their
synthesis, 1,3-dioxolane is allowed to react with acetyl bromide

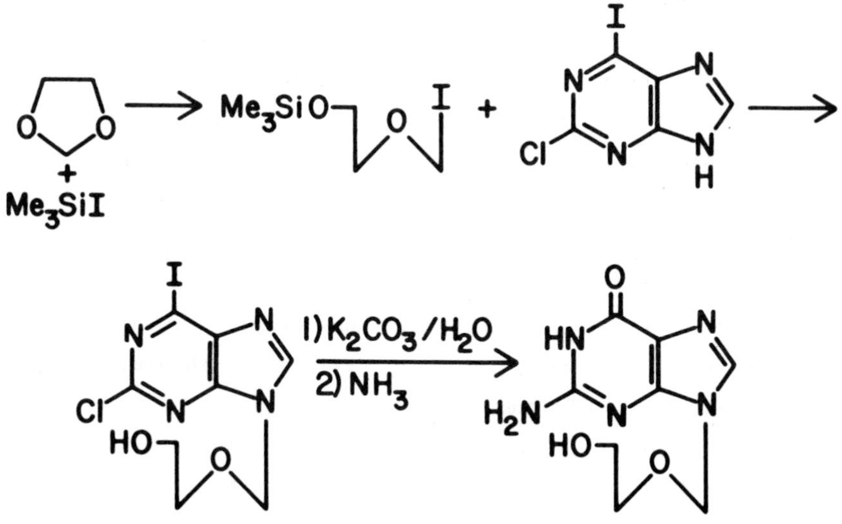

FIGURE 4. Synthesis of acyclovir (From Ref. 23).

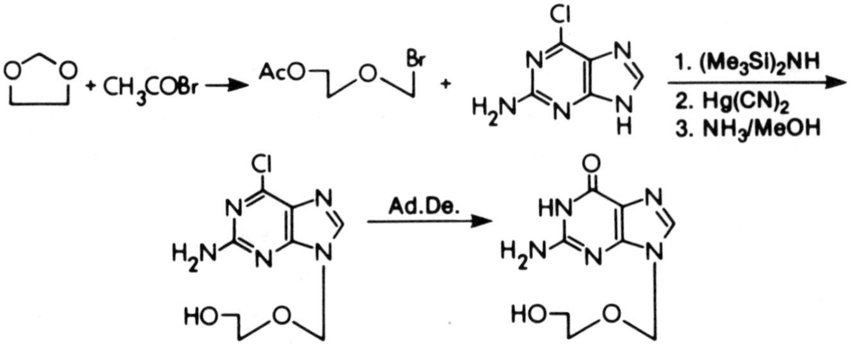

FIGURE 5.   Synthesis of acyclovir (From Ref. 24).

which generates acetoxyethyl bromomethyl ether.   Condensation
of the bromomethyl ether derivative with the trimethylsilyl
protected 2-amino-6-chloropurine gave, after deblocking, 2-amino-
6-chloro-9-(2-hydroxyethoxymethyl)purine.   Enzymatic hydrolysis
of the 6-chloro group with adenosine deaminase gave a good
yield of acyclovir (see Fig 5).   By a related sequence of
reactions using the appropriate trimethylsilylated pyrimidine
analogs and acetoxyethyl bromomethyl ether, Robins and Hatfield
prepared a variety of 1-(2-hydroxyethoxymethyl)pyrimidines
including the 5-hydrogen, 5-methyl, 5-fluoro, 5-chloro, 5-bromo,
5-nitro, and (E)-5-(2-bromovinyl)uracils (24).

        Once the potent antiviral activity of acyclovir was known,
a wide variety of nucleoside analogs were synthesized in which
variations were made in the heterocyclic moiety, the side chain
or in both.   Figure 6 shows the effect on antiviral activity

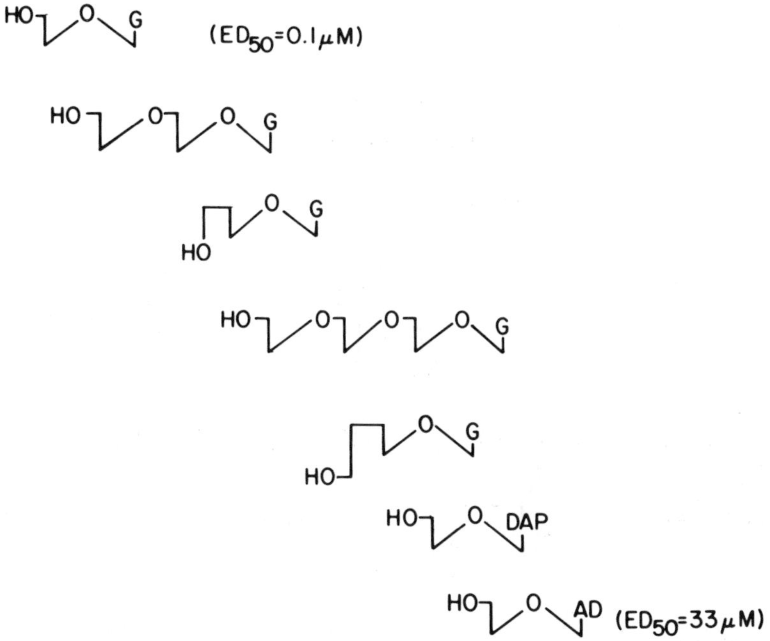

FIGURE 6.   Decreasing order of antiviral activity as a
function of structural modification of the side chain of
acyclovir(G=9-substituted guanine, DAP=9-substituted-2,6-diamino-
purine, and Ad=9-substituted adenine).

for certain of these structural changes. The compounds are
shown in decreasing order of antiviral activity.

Because acyclovir is an analog of guanosine, a variety of
structures related to guanosine were synthesized in our laboratories in which the ribose moiety of guanosine was modified in
various ways. Some of the compounds which were synthesized are

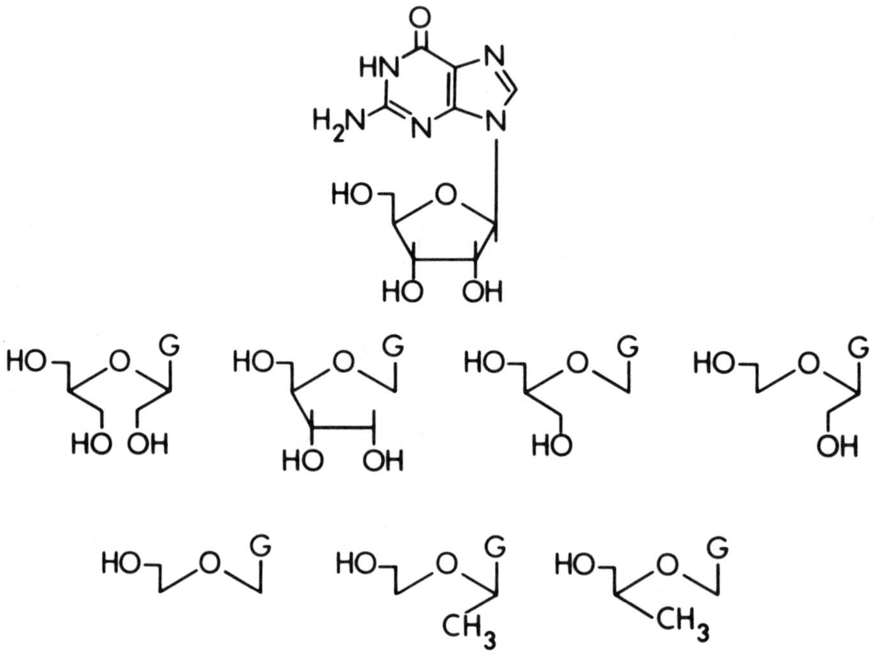

FIGURE 7.   Some structural modifications of guanosine(G=9-
substituted guanine).

shown in Fig. 7.   While all of these compounds possessed some
antiviral activity, the only compound of this group to possess
antiviral activity comparable to acyclovir was 9-[(2-hydroxy-1-
(hydroxymethyl)ethoxy)methyl]guanine.   The synthesis and biological evaluation of 9-[(2-hydroxy-1-(hydroxymethyl)ethoxy)methyl]-
guanine has recently been reported by three other laboratories
(25-28).

      Because a number of pyrimidine nucleoside analogs such as
TFT, IUDR, and AIU possess good antiviral activity, it was
reasoned that acyclic nucleoside analogs incorporating the
biologically important pyrimidine bases might have interesting

TABLE I.   Intracellular Concentration of Acyclovir and its
Phosphorylated Analogs[a]

$\underline{\text{pmole}/10^6 \text{ cells}}$

Form in Extract	Vero	HSV-1(H29) in Vero
Acyclovir	3.4	15.1
Acyclo-GMP	2.6	14.9
Acyclo-GDP	2.0	41.5
Acyclo-GTP	1.8	71.2
Total	9.8	142.7

[a]Vero cells were exposed to 100 µM acyclovir. Data taken
from Ref. 5.

antiviral activity. However, synthesis of a variety of acyclic
nucleoside pyrimidine analogs including 1-(2-hydroxyethoxy-
methyl)-5-(trifluoromethyl)uracil and 1-(2-aminoethoxymethyl)-5-
iodouracil revealed that these compounds exhibited little or no
activity against HSV-1 and against a range of other DNA and RNA
viruses (29,30).

Finally, the mechanism of antiviral action of acyclovir
appears to be similar to that of the other second generation
antiviral agents. The data in Table I were obtained in an
experiment in which HSV-1 infected vero cells and uninfected
vero cells were exposed to 100 µM acyclovir for a period of 7
hours after which the intracellular concentrations of acyclovir
and its phosphorylated analogs were measured (5). The total
intracellular amount of acyclovir and its phosphorylated analogs
was approximately 14 times higher in the HSV-1 infected cells

Howard J. Schaeffer

TABLE II. Apparent Kinetic Constants for DNA Polymerases[a]

Polymerase Source	Apparent K_m µM dGTP	Apparent K_i µM acyclo GTP
Hela	1.1	2.3
Vero	1.7	2.1
HSV-1/H29	0.38	0.08
HSV-1/KOS	0.97	0.55

[a]Data taken from Ref. 5.

than in the uninfected cells. More importantly, however, the
intracellular concentration of acyclovir triphosphate (acyclo-
GTP) was almost 40 times greater in the HSV-1 infected cells
than in the uninfected cells.

A second aspect of the mechanism of selective antiviral
activity of acyclovir involves the inhibition of DNA polymerase
by acyclovir triphosphate (5,31,32). The data in Table II show
the apparent Km values for some mammalian and HSV-1 DNA polymerases and the apparent Ki values of acyclovir triphosphate
(acyclo GTP) for these same enzymes. These results show that
acyclovir triphosphate is a significantly more potent inhibitor
of the virally-induced DNA polymerase than it is of the host
α-DNA polymerase. Thus, the selective antiviral effect of
acyclovir may be explained by the fact than it is a much better
substrate of herpes-induced thymidine kinase than it is of
host-cell thymidine kinase. This results in a higher intracellular concentration of the monophosphate and ultimately of
acyclovir triphosphate in herpes-infected cells compared to the
uninfected cells. In addition, acyclovir triphosphate is a
more potent inhibitor of herpes-induced DNA polymerase than it
is of mammalian α-DNA polymerase. These combined effects result
in a compound with potent and selective antiviral activity.

REFERENCES

1. Davies, W. L., Grunert, R. R., Haff, R. F., McGahen, J. W., Neumayer, E. M., Paulshock, M., Watts, J. C., Wood, T. R., Hermann, E. C., and Hoffmann, E. C., Science, 144, 862 (1964).

2. Prusoff, W. H., Biochim. Biophys. Acta 32, 295 (1959).

3. Kaufman, H. E. and Heidelberger, C., Science 145, 585 (1964).

4. Schaeffer, H. J., Beauchamp, L., de Miranda, P., Elion, G. B., Bauer, D. J., and Collins, P., Nature 272, 583 (1978).

5. Elion, G. B., Furman, P. A., Fyfe, J. A., de Miranda, P., Beauchamp, L., and Schaeffer, H. J., Proc. Natl. Acad. Sci. U.S.A. 74, 5716 (1977).

6. Drach, G. C., Ann. Rep. Med. Chem. 13, 139 (1978).

7. Prusoff, W. H., Chen, M. S., Fischer, P. H., Lin, T-S., Shiau, G. T., Schinazi, R. F., and Walker, J., Pharmacol. Therap. 7, 1 (1979).

8. Heidelberger, C. and King, D. H., Pharmacol. Therap. 6, 427 (1979).

9. Lin, T-S., Neenan, J. P., Cheng, Y-C., Prusoff, W. H., and Ward, D. C., J. Med. Chem. 19, 495 (1976).

10. Chen, M. S., Ward, D. C., and Prusoff, W. H., J. Biol. Chem. 251, 4833 (1976).

11. Sidwell, R. W., Robins, R. K., and Hillyard, I. W., Pharmacol. Therap. 6, 123 (1979).

12. Gentry, G., McGowan, J., Barnett, J., Nevins, R., and Allen, G., Adv. Ophthalmol. 38, 164 (1979).

13. DeClercq, E., Descamps, J., DeSomer, P., Barr, P. J., Jones, A. S., and Walker, R. T., Proc. Natl. Acad. Sci. U.S.A. 76, 2947 (1979).

14. Watanabe, K. A., Reichman, U., Hirota, K., Lopez, C., and Fox, J. J., J. Med. Chem. 22, 21 (1979).

15. Fox, J. J., Lopez, C., and Watanabe, K. A., Antiviral Chemo., 219 (1981).

16. Kit, S., Pharmacol. Therap. 4, 501 (1979).

17. Fischer, P. H., Chen, M. S., and Prusoff, W. H., Biochem. Biophys. Acta 606, 236 (1980).

18. Lopez, C., Watanbe, K. A., and Fox, J. J., Antimicrob. Agents Chemother. 17, 803 (1980).

19. Cheng, Y-C., Dutschman, G., Fox, J. J., Watanabe, K. A., and Machida, H., Antimicrob. Agents Chemother. 20, 420 (1981).

20. Allaudeen, H. S., Kozarich, J. W., Bertino, and De Clercq, E., Proc. Natl. Acad. Sci. U.S.A. 78, 2698 (1981).

21. De Clercq, E., Acta Microbiol. Acad. Sci. Hung. 28, 289 (1981).

22. Schaeffer, H. J., Gurwara, S., Vince, R., and Bittner, S., J. Med. Chem. 14, 367 (1971).

23. Barrio, J. R., Bryant, J. D., and Keyser, G. E., J. Med. Chem. 23, 572 (1980).

24. Robins, M. J. and Hatfield, P. W., Can. J. Chem. 60, 547 (1982).

25. Smith, K. O., Galloway, K. S., Kennell, W. L., Ogilvie, K. K., and Radatus, B. K., Antimicrob. Agents Chemother. 22, 55 (1982).

26. Ogilvie, K. K., U.S. Patent No. 4,347, 360 (1982).

27. Verheyden, J. P. and Martin, J. C., U.S. Patent No. 4,355, 032 (1982).

28. Ashton, W. T., Karkas, J. D., Field, A. K., and Tolman, R. L., Biochem. Biophys. Res. Comm. 108, 1716 (1982).

29. Kelley, J. L., Kelsey, J. E., Hall, W. R., Krochmal, M. P., and Schaeffer, H. J., J. Med. Chem. 24, 753 (1981).

30. Kelley, J. L., Krochmal, M. P., and Schaeffer, H. J., J. Med. Chem. 24, 472 (1981).

31. Furman, P. A., St. Clair, M. H., Fyfe, J. A., Rideout, J.
    L., Keller, P. M., and Elion, G. B., J. Virol. 32, 72
    (1979).
32. Elion, G. B., Adv. Enzyme Regul. 18, 53 (1980).

THE CHEMISTRY AND BIOLOGY OF
NUCLEOSIDES OF PURINES AND RING ANALOGS

John A. Montgomery

Organic Chemistry Research Department
Southern Research Institute
Birmingham, Alabama

I.   CHEMISTRY

2-Chloro- ($\underline{1}$, X = Cl)[1] and 2-bromoadenosine ($\underline{1}$, X = Br) are best prepared by ammonolysis of the corresponding 2,6-dihalo-purine ribonucleosides usually as the triacetates, conveniently synthesized by the fusion[2] of the dihalopurines with tetraacetyl ribofuranose,[3] although other methods are available.[4] Ammonolysis readily displaces the 6-halo group with concomitant removal of the O-acetyl groups. Replacement of the 6-halo group by the amino function deactivates the 2-halo group which is then not displaced by ammonia under the usual conditions.[5] This same approach can be used for the 3'-deoxyadenosine ($\underline{2}$) and the 9-β-$\underline{D}$-xylofuranosyl-adenines ($\underline{3}$).[6] The fusion, and other methods, for the attachment of 2'-deoxyribose are complicated by lack of the stereochemical control of the 2-acyloxy function of the sugar resulting in the production of approximately equal amounts of the α- and β-anomers, which after conversion by ammonolysis to the 2'-deoxy-adenosines ($\underline{4}$), can be separated chromatographically.[7] When applied to arabinose the fusion reaction gives predominantly the

Nucleosides, Nucleotides,
and Their Biological Applications

19

α-anomer, since stereochemical control by the orthoester ion of this sugar directs the entering purine base to attack from the α- and not the β-side.[6,8] This difficulty is overcome by use of 2,3,5-tri-O-benzyl-β-D-arabinofuranosyl chloride in a nonpolar solvent in the presence of molecular sieves.[9] Under these S_N2 conditions, displacement of the chloro group occurs almost exclusively from the β-side, but after replacement of the 6-halo groups, the benzyl groups must be removed catalytically[10] or by means of boron trichloride[11] to give 5.[12]

The preparation of the 8-chloro- and bromoadenine nucleosides is much simpler, because both halogens can be introduced directly into the corresponding adenine nucleosides (6 and 7) with or without protecting groups on the sugar hydroxyls, except in the

case of 9-β-D-arabinofuranosyl adenine, which, because of the
facile displacement of the 8-halogen by the cis 2'-hydroxyl gives
only the anhydronucleoside.[17,18] 8-Chloro-2'-deoxyadenosine (8,
X = Cl, R' = H), not previously reported, was prepared from 2'-deoxy-
adenosine (6, R' = H) in 28% yield by treatment with t-butyl
hypochlorite in DMF.

Synthesis of fluoropurine nucleosides is more difficult, since
the methods described above are not applicable to their
preparation.   2-Fluoroadenosine (1, X = F) was originally
prepared by a modification of the well-known Schiemann
reaction.[19] The original synthesis was vastly improved by using
2-aminoadenine nucleosides with their sugar hydroxyls blocked by
acyl groups.[6] These compounds could be conveniently prepared by
reaction of the 2,6-dihalo purine nucleosides, prepared as
described above, with sodium azide followed by reduction to the
diaminopurine nucleosides, both of these steps proceeding in high
yield.[6] Because of the stability of the O-benzyl groups to base
treatment, 9-β-D-arabinofuranosyl-2-fluoroadenine (5, X = F, R =
OH) can be more conveniently prepared from 2,6-diacetamidopurine.[12]

The preparation of 8-aza-2-fluoroadenosine (18) presented a
somewhat different problem in that 8-aza-2,6-dichloropurine is
too unstable[20] to be used in nucleoside synthesis. Condensation
of 8-aza-2,6-bis(methylthio)purine (10)[8] with 1-chloro-2,3,5-
tri-O-acetyl-D-ribofuranose (11) in toluene in the presence of
molecular sieves gave the desired 9-isomer (12) in 70-90% yield
along with a minor amount of the 8-isomer (13). Although
separation of 12 and 13 was not convenient, the 8-azaadenine
nucleosides (15 and 16), prepared by treatment of the mixture of
12 and 13 with ethanolic ammonia, could be separated chromatographically. Oxidation of 15 with m-chloroperbenzoic acid in
methanol afforded 14 in 80-90% yield. In later preparations 14
crystallized from the reaction mixture from the oxidation of a
mixture of 15 and 16. The methylsulfonyl group at C-2 of 14 was
easily displaced by ethanolic ammonia (21 hours at room

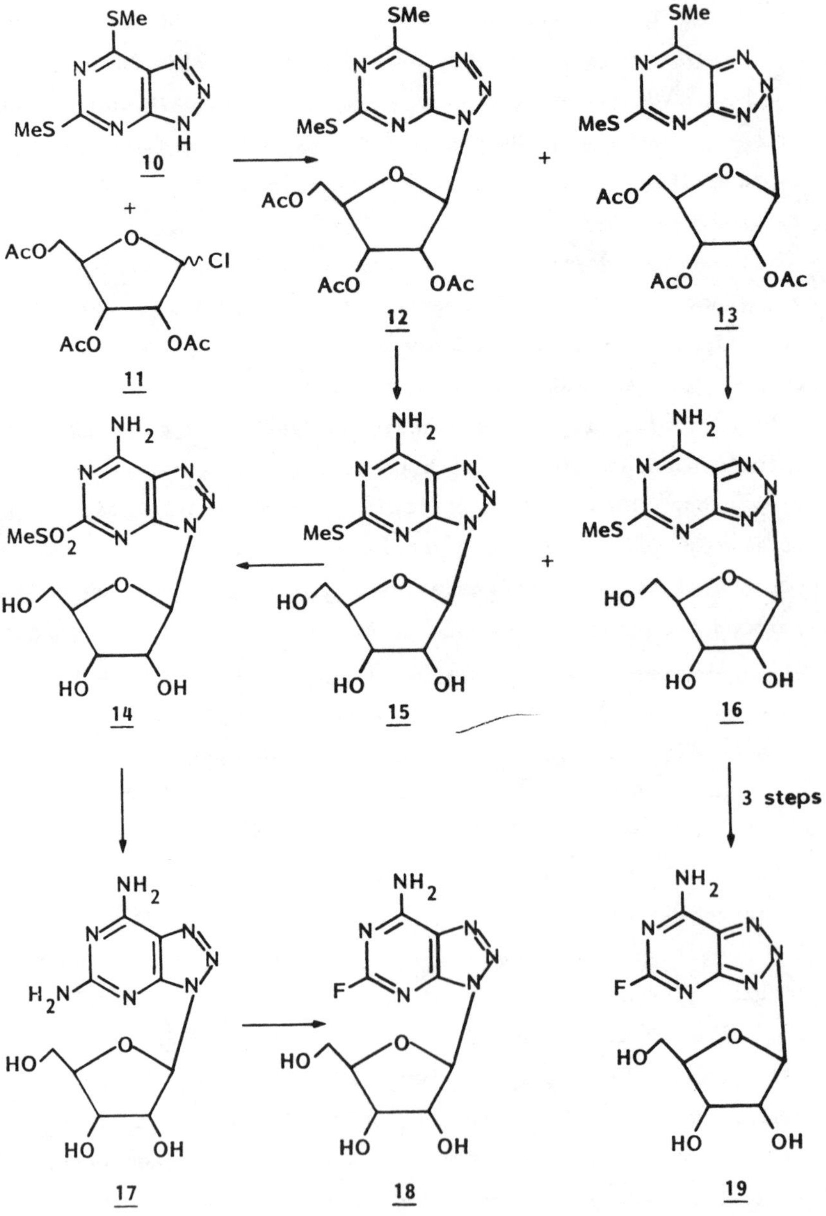
SMe
N
N
N
MeS
N
N
H
10
+
AcO
O
Cl
AcO
OAc
11
SMe
MeS
AcO
AcO
OAc
12
+
SMe
MeS
AcO
AcO
OAc
13
NH2
MeSO2
HO
HO
OH
14
NH2
MeS
HO
HO
OH
15
+
NH2
MeS
HO
HO
OH
16
3 steps
NH2
H2N
HO
HO
OH
17
NH2
F
HO
HO
OH
18
NH2
F
HO
HO
OH
19

temperature) to give 2-amino-8-azaadenosine (17). Treatment of 17 with potassium nitrite in fluoroboric acid followed by neutralization with potassium hydroxide allowed removal of most of slightly soluble potassium fluoborate. Absorption on resin beads permitted removal of the remaining inorganic salts by aqueous wash. Essentially pure 8-aza-2-fluoroadenosine (18) was then eluted from the beads with ethanol. The 8-isomer (19) was prepared by the same sequence from 13 prepared by the fusion reaction.

Although the synthesis of 8-fluoroadenosine (22) from 8-aminoadenosine triacetate (20) by the improved procedure described above has been claimed,[21] we have been unable to repeat the work described, in agreement with the conclusion of Kobayashi et al. who prepared the triacetate of 8-fluoroadenosine (21) by a "naked fluoride" displacement reaction, but were unable to convert it to 8-fluoroadenosine (22) itself.[22] We have now prepared the

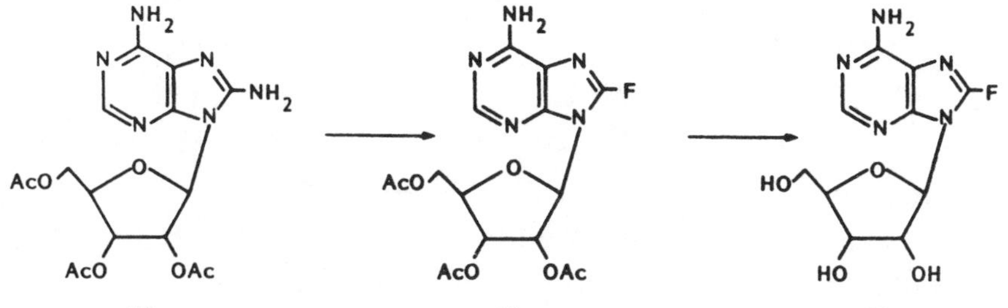

20

21

22

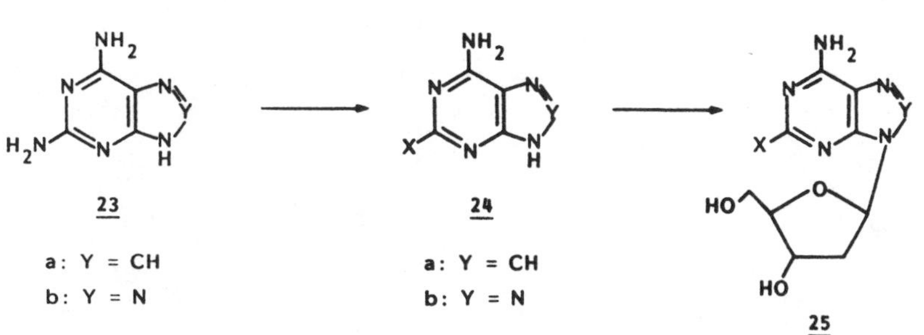

23

a: Y = CH
b: Y = N

24

a: Y = CH
b: Y = N

25

a: Y = CH
b: Y = N

8-fluoroadenosine triacetate (21) by diazotization with inorganic or t-butyl nitrite of 8-aminoadenosine triacetate (20) in a mixture of HF in pyridine.[23] M. J. Robins et al. have independently applied this method to the synthesis of 2-fluoro-purine nucleosides.[24] The major product from 8-aminoadenosine triacetate (20) was adenosine triacetate, an unexpected and unexplained result. Treatment of 21 with ethanolic ammonia at ambient temperature gave 8-fluoroadenosine in 5-10% yield from 20. This nucleoside has now been fully characterized: by U.V., [1]H-nmr, ^{13}C-nmr, and mass spectroscopy, and chromatographically. Its biologic properties differ from those reported.[25]

The limitation of chemical methods for the preparation of 2'-deoxyribonucleosides was discussed above. Recently described methods[26-28] for the conversion of ribonucleosides to 2'-deoxy-ribonucleosides in good yield, unfortunately, do not appear to be applicable to halopurine nucleosides. The 2-haloadenines can, however, be facilely converted to their 2'-deoxyribonucleosides (25a) enzymatically;[29,30] but, in the case of 2-fluoroadenine (24a, X = F), this approach necessitated the development of a better synthesis of the purine (24a, X = F), which was accomplished by means of the pyridine-HF procedure.[23] 2-Fluoro-8-azapurine (24b, X = F) was also prepared in this way and converted to its 2'-deoxyribonucleoside (25b, X = F). 9-β-D-Arabino-furanosyl-2-fluoroadenine (5, X = F) has been prepared enzymatically as well,[31] in a somewhat different manner.[32]

Application of the enzymatic approach to the preparation of 2'-deoxyribonucleosides from 8-haloadenines (26) led to complications, in that attachment occurred exclusively, in the case of the trifluoromethyl compound 26 (X = CF$_3$), at N-3 (29, X = CF$_3$) rather than N-9 and in part, in the case of the chloro (26, X = Cl) and bromo (26, X = Br) compounds.[33] Also the deoxyribosyltransferase was able to convert the N-3 nucleosides (29) partially to N-9 (28) and, to a lesser extent, the N-9 (28) nucleosides to N-3 (29). That these anomalous reactions result from electronic rather than steric factors seems clear from the fact that no N-3 nucleoside is formed from 8-methyladenine (X = CH$_3$). In any event, the yields are not good, and the mixtures must be separated, making this method impractical for the preparation of the 8-substituted adenine 2'-deoxyribonucleosides.

2-Fluoroadenosine ($\underline{1}$, X = F) was converted to its 2',3'-$\underline{O}$-
ethoxymethylidene derivative ($\underline{30}$), which was treated with tosyl
chloride in pyridine to prepare the 5'-$\underline{O}$-tosyl derivative $\underline{31}$.
Treatment of $\underline{31}$ with NaI in acetone gave 5'-deoxy-2',3'-$\underline{O}$-
ethoxylidene-2-fluoro-5'-iodoadenosine ($\underline{32}$), which was converted
to 5'-deoxy-2-fluoro-5'-iodoadenosine   ($\underline{35}$)   by   mild   acid.
Treatment of either $\underline{31}$ or $\underline{32}$ with sodium methylmercaptide in DMF
followed by aqueous acid gave 5'-deoxy-2-fluoro-5'-methylthio-
adenosine ($\underline{34}$).

The synthesis of the carbocyclic analog of 3-deazaadenosine
($\underline{38}$) by a sequence similar to that used for the corresponding
purine, but starting with 2,4-dichloro-3-nitropyridine $\underline{36}$ has
recently been described.[34]   A different sequence employing the
reaction of 4,6-dichloro-5-(2,2-diethoxyethyl)pyrimidine ($\underline{39}$)[35]
with   ($\underline{+}$)(1,4/2,3)-4-amino-2,3-dihydroxy-1-cyclopentanemethanol
($\underline{37}$)[36,37] was used for the preparation of the carbocyclic analog
($\underline{42}$) of 7-deazaadenosine (tubercidin).

II. BIOLOGY

It is now well established that halogenation of adenine nucleosides at C-2 greatly reduces their deamination by adenosine deaminase (ADA),[38-41] but the other consequences of this halogenation depend on both the halogen and sugar moiety. Thus, 2-fluoro-, 2-chloro-, and 2-bromoadenosine (1) are all quite resistant to ADA (Table 1), but only 2-fluoroadenosine shows a high level of cytotoxicity (Table 2). The decrease in toxicity (F>Cl>Br) can be correlated with the ability of these nucleosides to serve as substrates for another adenosine metabolizing enzyme, adenosine kinase, since 2-fluoroadenosine is an excellent substrate for this enzyme, whereas 2-chloro- and 2-bromoadenosine are poor substrates.[44,45] Unfortunately, none of these analogs has shown significant selective cytotoxicity, as is evident from their lack of in vivo anticancer activity. The basis of their

Table 1.   Kinetic Constants of 2-Haloadenine Nucleosides as Substrates for Adenosine Deaminase[a]

Nucleoside	K_m (μM)	V_{max} (μmoles/min/mg)	Relative Substrate Efficiency[b]
Ado	29	435	100
2-F-Ado	81	0.78	0.06
2-F-dAdo	71	1.9	0.20
2-F-ara-A	220	0.075	0.002
2-Cl-Ado	110	0.06	0.004
2-Cl-dAdo	47	0.06	0.009
2-Br-Ado	63	0.01	0.001
2-Br-dAdo	57	0.009	0.001

[a]Sigma, calf intestine.   [b]V_{max}/K_m X 6.67.

Table 2.   Cytotoxicity of the 2-Haloadenosines[a]

Adenosine	IC_{50}, μM [b]	Degree of Resistance[c]		
		APRT⁻	AK⁻	APRT⁻/AK⁻
2-Fluoro-	0.02	20	1-2	>2000
2-Chloro-	7	1	>10	>10
2-Bromo-	120	1		

[a]To Human epidermoid carcinoma (H.Ep. #2) cells
in culture.   [b]The concentration required to inhibit
the growth of treated cells to 50% of untreated
controls.[c]Ratio of the IC_{50} for the mutant cell
line to the IC_{50} for the parent line: APRT⁻ cells
selected for resistance to 2-fluoroadenine, AK⁻
cells selected for resistance to 6-(methylthio)-
purine ribonucleoside, APRT⁻/AK⁻ cells selected
for resistance to 2-fluoroadenine and then 2-fluoro-
adenine. Data from References 42 and 43.

cytotoxicity has not been   established but 2-fluoroadenosine (1,
X = F) and 2-fluoroadenine are metabolized to the mono-,[44] di-,
and triphosphates (43),[46] to the 3',5'-cyclic monophosphate
(44),[47-49] to 2-fluoro-S-adenosylhomocysteine (46)[50] and to 2-
fluoro-S-adenosylmethionine (47).[51] There is also evidence that
2-fluoroadenosine may be incorporated into RNA.[46]

    The knowledge that incorporation of the 2-fluoro group into
adenosine greatly reduces the rate of deamination (0.002 the
V_{max} of adenosine), although only raising the K_m about 3-
fold, led to the preparation of other 2-fluoroadenine nucleosides[6] and 2-fluoro-N-substituted adenine[52] nucleosides with the
hope that the anticancer activity of compounds like 9-β-D-ara-
binofuranosyladenine (ara-A, 5, X = H), 9-β-D-xylofuranosyl-
adenine (3, X = H), and N-hydroxyadenosine could be increased.
Other investigators led by Le Page[53] were able to enhance the
activity of adenine nucleosides[54,55] by means of adenosine
deaminase inhibitors, but, unfortunately, these ADA inhibitors
cause other effects also.[56,57] The success of our approach

(introduction of F at C-2) is apparent from the increased
toxicity of these 2-fluoropurine nucleosides, although a notable
exception appears to be the 2-fluoro analog of 9-β-D-
xylofuranosyladenine (3, X = F) (Table 3). These results were
extended to the in vivo evaluation of 9-β-D-arabinofuranosyl-2-
fluoroadenine (F-ara-A) (5, X = F) against the murine leukemias
L1210 and P388 on a variety of schedules.[58] Administration of F-
ara-A daily for nine days is as effective as any other schedule
tried and, on this schedule, is as effective as ara-A given with
2'-deoxycoformycin (DCF) on a more complex schedule (every 3
hours for 24 hours on days 1, 5, and 9.[59]

Table 3. Cytotoxicity of 2-Fluoroadenine and Some
          of its Nucleosides[a]

Compound	IC_{50} (μM)	Degree of Resistance		
		APRT⁻	AK⁻	APRT⁻/AK⁻
2-Fluoroadenine	0.03	> 2000	1	> 2000
2-Fluoroadenosine	0.02	20	1-2	> 2000
2'-Deoxy-2-fluoroadenosine	0.2	2	1	2
α-Anomer	6.5			
3'-Deoxy-2-fluoroadenosine	2.0	40	1	44
3'-Deoxyadenosine	80	1	1	
2-Fluoro-9-β-D-xylofuranosyladenine	40			
9-β-D-xylofuranosyladenine	30			
9-β-D-arabinofuranosyl-2-fluoroadenine	8.0			
α-Anomer	40			
9-β-D-arabinofuranosyladenine	> 400			
α-Anomer	40		> 80	
5'-Deoxy-2-fluoroadenosine	0.1	> 100		
5'-Deoxy-5'-ethylthio-2-fluoroadenosine	0.2			> 400

[a]See footnotes to Table 2. Data from References 6, 69, and 70.

The success of F-ara-A prompted an investigation of the activity of 2'-deoxy-2-fluoroadenosine (**4**, X = F) and 2-chloro-2'-deoxyadenosine (**4**, X = Cl) against leukemia L1210 (Table 4). Both compounds proved to be curative also, but like ara-A plus DCF, only when given every 3 hours for 24 hours on days 1,5, and 9.[29] Given daily, 2-chloro-2'-deoxyadenosine was less effective, presumably, because lethal levels of the triphosphate (**50**, X = F, Cl) are not maintained in the leukemia cells long enough to kill all of the kinetically heterogeneous population as each cell progresses through the S-phase of the cell cycle.[60] Conversion of **4** (X = Cl) to the 5'-O-nonanoyl ester, however, gave a depot form that is curative on a daily schedule. In contrast, daily administration of F-ara-A itself (or as the 5'-phosphate) was shown to be sufficient to maintain effective levels of its triphosphate (51, X = F),[58] a potent inhibitor of both nucleoside diphosphate reductase and DNA polymerase.[61] Extension of these studies to 2-bromo-2'-deoxyadenosine (4, X = Br) showed that it, too, is curative in the L1210 system, on the every three-hour schedule. Furthermore, in contrast to the results with the ribonucleosides, the 2-chloro- and 2-bromo-2'-deoxyadenosines (4, X = Cl and Br) are more cytotoxic to cultured cells and more toxic to mice than the 2-fluoro compound (4, X = F) (Table 5). These 2-halo derivatives (1) of 2'-deoxyadenosine appear to be similar in their metabolism and their mechanism of action to F-ara-A, but definitive studies have not been carried out. The cytotoxicity of the fluoro and chloro compounds are reversed by 2'-deoxycytidine

Table 4.   In Vivo Activity of the 2-Halo-2'-deoxyadenosines vs. Leukemia L1210

Nucleoside	Dosage[a] (mg/kg /dose)	Schedule[b]	% ILS[c]	Cures /Treated
2'-Deoxyadenosine	660	A	10	0/8
2'-Deoxyadenosine + 2'-Deoxycoformycin	150 + 0.05	A	32	0/10
2-Chloro-2'-deoxyadenosine	50	B	81	0/6
	15	A	175	5/10
2-Chloro-2'-deoxyadenosine 5'-nonanoate	50	C	244	4/6
2'-Deoxy-2-fluoroadenosine	40	A	100	3/10
2-Bromo-2'-deoxyadenosine	40	A	125	3/6

[a]Mg/kg/dose (at less than or equal to the lethal dose for 10% of the animals (autopsied for cause of death)).   [b]A = every three hours for 24 hours on days 1,5, and 9; B = daily, days 1-6; C = daily, days 1-9.   [c]Percent increase in lifespan.

NH$_2$ NH$_2$

Ribonucleotides

dCK

$^-$H$_3$O$_9$P$_3$O

DNA

48: R = H
49: R = OH

4: R = H
5: R = OH

50: R = H
51: R = OH

and cells lacking 2'-deoxycytidine kinase are resistant to them, indicating that they are phosphorylated by this enzyme only, as has been established for F-ara-A.[59] They are further converted to their triphosphates (50), which preferentially inhibit DNA synthesis.[29] Thus, the weight of evidence supports the similarity of these compounds to the well-studied F-ara-A.[58,59,61-65] Even less work has been done on 9-β-D-arabinofuranosyl-2-chloroadenine, but it is logical to assume that it, too, is behaving like F-ara-A.

The impressive anticancer activity of the 2-haloadenine 2'-deoxyribonucleosides caused us to reinvestigate the biologic activity of the 8-haloadenine nucleosides (8 and 9, X = F, Cl, Br) as well. These compounds, with the exception of 8-fluoroadenosine (22), are also highly resistant to ADA (Table 6), but, in contrast to the 2-halo nucleosides, little difference is observed in the cytotoxicity of the 8-fluoro-, 8-chloro-, and 8-bromoadenosines (9) (Table 5). In this case the increasing size of the halogen does not appear to interfere, presumably because these compounds are all readily phosphorylated by adenosine kinase.[45] In the 2'-deoxyribose series the opposite is true. 2-Chloro- and 2-bromo-2'-deoxyadenosine (4, X = Cl, Br) are more cytotoxic than the 2-fluoro compound, (4, X = F) but the 8-chloro and the 8-bromo compounds (8, X = Cl, Br) are only cytotoxic at high levels (Table 5). In all cases the bromoadenine nucleosides (1, 4, 8, and

Table 5. Cytotoxicity of the Haloadenines and Some of their Nucleosides

IC_{50} Values[a]

R =	H	HO (ribose) HO OH	HO (2'-deoxy) HO	HO (arabino) HO HO
Y = H, X = H	100	1.0^b	2.6^b	ca. 5^b
F	0.03	0.02	0.2	8
Cl	10	7.0	0.02	ca. 3
Br	70	120	0.02	--
X = H, Y = H	100	1.0^b	2.6^b	ca. 5^b
F	--	2.0	--	--
Cl	> 60	0.03	>100	--
Br	> 60	2.0	400	--

[a] μMolar concentration of compound required to inhibit growth of treated H.Ep. #2 cells to 50% of controls. [b] In the presence of 2'-deoxycoformycin.

9, X = Br) are most resistant to deamination (Tables 1 and 6), but the differences in the rate of deamination, except in the case of 8-fluoroadenosine (**22**) already referred to, would appear to be, at most, a minor determinant in activity.

The cytotoxicity of the 2-haloadenosines (**1**), as stated above, appears to relate to their ability to serve as substrates for adenosine kinase as determined by the size of the halogen. Only limited quantitative data is available for the other nucleosides,[66] but the evidence given above indicates that both the 2'-deoxyribonucleosides (**4**) and the arabinonucleosides (**5**) are phosphorylated by 2'-deoxycytidine kinase and not by adenosine kinase (ara-A is phosphorylated by both). Thus, there appears to be limited bulk tolerance at C-2 but not at C-8 of the

Table 6.   Kinetic Constants of 8-Haloadenine
           Nucleosides as Substrates for
           Adenosine Deaminase[a]

Nucleoside	K_m (μM)	V_{max} (μmoles/min/mg)	Relative Substrate Efficiency[b]
Ado	29	435	100
8-F-Ado	1000	233	1.5
8-Cl-Ado	830	7.6	0.06
8-Cl-dAdo	670	18	0.20
8-Br-Ado	250	0.04	0.001
8-Br-dAdo	400	0.12	0.002

[a]Sigma, calf intestine.   [b]V_{max}/K_m X 6.67.

purine moiety in the case of adenosine kinase and the opposite
appears to be true with 2'-deoxycytidine kinase.   Work is in
progress to document this conclusion.

Among the haloadenine nucleosides, 8-fluoroadenosine (22)
is an anomaly in that it is deaminated at a significant rate,
although much slower than adenosine (Table 6).   The potent ADA
inhibitor 2'-deoxycoformycin, however, does not affect its cytotoxicity to H.Ep. #2 or L1210 cells, except in the case of L1210
cells deficient in adenosine kinase (L1210/MeMPR and
L1210/MP/MeMPR), which are highly resistant to 8-fluoroadenosine
in its presence and only slightly resistant in its absence,
indicating the necessity for conversion to 8-fluoroinosine for
toxicity to those cell lines (Table 7).   These results indicate
that 8-fluoroadenosine (22) is normally phosphorylated to 8-
fluoroadenylic acid (53) by adenosine kinase, but,   an
alternative pathway to a cytotoxic nucleotide exists via
deamination to 8-fluoroinosine 52.   Phosphorolysis to 8-fluoro-
hypoxanthine (54) may occur but conversion of 54 to 8-fluoroino-

Table 7. Cytotoxicity of 8-Fluoroadenosine[a]

	IC$_{50}$ (μM)			
		L1210		
	H.Ep. #2	/0	/AK^{-}[b]	/HPRT^{-}/AK^{-}[c]
8-Fluoroadenosine	4	0.5	4	5
8-Fluoroadenosine + DCF[d]	~1	0.3	> 100	> 40

[a]See Footnotes to Table 2. [b]Resistant to 6-methylthio-purine ribonucleoside. [c]Resistant to 6-mercaptopurine and 6-methylthiopurine ribonucleoside. [d]DCF = 2'-deoxycoformycin.

sinic acid (55) via HPRT seems unlikely in view of the results with the double mutant. If 55 is formed by some other route it may be further metabolized to either 8-fluoroadenylic (53) or 8-fluoro-guanylic (56) acid or both. In cells having adenosine kinase it is also possible that 52 may be phosphorylated as well (to 55). Moreover, it is possible that 52 is in itself cytotoxic, but that possibility seems less likely.

Table 8.   Kinetic Constants of 8-Azaadenosines as Substrates
           for Adenosine Deaminase[a]

Nucleoside	K_m (μM)	V_{max} (μmoles/min/mg)	Relative Substrate Efficiency[b]
Ado	29	435	100
8-Aza-Ado	250	1500	40
2-F-8-Aza-Ado	250	5	0.20

[a]Sigma, calf intestine.   [b]V_{max}/K_m X 6.67.

Table 9.   Cytotoxicity of 8-Azaadenine and Derivatives[a]

Compound	IC_{50} H.Ep. #2	L1210/0	L1210/AK⁻
8-Azaadenine	20		
8-Aza-2-fluoroadenine	200		
8-Azaadenosine	0.7		
8-Azaadenosine + DCF		< 0.04	> 4
8-Aza-2-fluoroadenosine	3		
8-Aza-2-fluoroadenosine + DCF	3		
8-Aza-2'-deoxyadenosine	6		
8-Aza-2'-deoxyadenosine + DCF	> 20		
8-Aza-2'-deoxy-2-fluoroadenosine	~ 30		
8-Aza-2-fluoro-8-D-ribofuranosyladenine	> 70		

[a]See footnotes to Table 7.

8-Azaadenosine is a good substrate for adenosine deaminase It has a V_{max} three and one-half times that of adenosine (Table 8) so that it is rapidly deaminated in whole cells, even though it is also phosphorylated to 8-azaadenylic acid.[67] Introduction of a fluorine at C-2 of this nucleoside to give 18 did not affect the K_m but reduced the V_{max} to 1/300 that of 8-azaadenosine. Unfortunately, it also reduced its cytotoxicity and antileukemic activity. The cytotoxicity of 8-aza-2'-deoxyadenosine (25b, X = H) was reduced the same relative amount by the introduction of fluorine at C-2 to give 25b (X = F) (Table 9).

5'-Deoxy-2-fluoroadenosine (57, R = H, X = F)[68] and 5'-deoxy-5'-ethylthio-2-fluoroadenosine (57, R = EtS, X = F),[69] two nucleosides that cannot be phosphorylated, are cytotoxic to H.Ep. #2 cells in culture (Table 3) but have no activity against L1210 leukemia in vivo, although they were much less toxic to mice than 2-fluoroadenosine. Their lack of activity against cells deficient in adenine phosphoribosyltransferase indicates that cleavage to 2-fluoroadenine is a necessary step in their activation.[68,69]

Table 10.   Cytotoxicity of Some Analogs of 5'-Deoxy-5'-Methylthioadenosine[a]

Nucleoside	IC_{50} (μM)
5'-Deoxy-5'-iodo-2-fluoroadenosine	0.2 ± 0.1 (4)[b]
5'-Deoxy-5'-iodoadenosine	146.0 ± 6.0 (2)
5'-Deoxy-5'-ethylthio-2-fluoroadenosine	0.4 ± 0.2 (2)
5'-Deoxy-5'-ethylthioadenosine	approx. 85 (1)
2-Fluoroadenine	0.7 ± 0.02 (2)

[a]To HCT-15 human colon carcinoma cells; see Footnotes to Table 2. [b]Number of experiments.(Data supplied by T. M. Savarese and R. E. Parks, Jr.)

5'-Deoxy-5'-ethylthio-2-fluoroadenosine (57, R = EtS, X = F) has been found   to be cleaved by 5'-deoxy-5'-methylthio adenosine phosphorylase (MTPase),[70] an enzyme present in H.Ep. #2 cells but absent in L1210 cells.   These compounds are then prodrug forms of 2-fluoroadenine (58, X = F), which are less toxic to mice than the parent, and, as such should, show activity against neoplasms containing MTPase.   5'-Deoxy-5'-ethylthioadenosine (57, R = EtS, X = H) and 5'-deoxy-5'-iodoadenosine (57, R = I, X = H) are also cytotoxic to cells with MTPase such as HCT-15 human colon carcinoma (Table 10), but in these cases the cytotoxicity must be due to the sugar moiety, which may be converted to ethionine (60, R = EtS) and 2-iodoalanine (60, R = I) by the enzymes that convert 5-deoxy-5-methylthioribose to methionine.[71,72] These amino acids could be the cytotoxic agents.   5'-Deoxy-2-fluoro-5'-iodoadenosine (35) as well as 5'-deoxy-5'-ethylthio-2-fluoroadenosine (57, R = EtS, X = F) are cleaved by MTA phosphorylase to give two cytotoxic agents, and both these nucleosides are more cytotoxic than 2-fluoroadenine, but the differences are not great.   5'-Deoxy-5'-methylthio-2-fluoroadenosine (34) is also cytotoxic to H.Ep. #2 cells but, since an accurate ED_{50} has not been determined, the contribution of the sugar moieties of the 5'-ethylthio and 5'-iodo compounds cannot yet be made.

3-Deazaadenosine (62)[73,74] is another ADA resistant adenosine analog but, in contrast to 2-fluoroadenosine, it is also poorly phosphorylated and is less cytotoxic than 3-deazaadenine. It is an alternative substrate for, and consequently a good competitive inhibitor of, adenosylhomocysteinase[75], an interaction that is thought to be responsible for its antiviral activity, through a build up of 64 (X = O) and adenosylhomocysteine.[76,77] The carbocyclic analog of adenosine, (±)-9-[(1α,3β,3β,4α)-2,3-dihydroxy-4-hydroxymethyl)cyclopentyl]adenine (63),[78] by far the most potent known inhibitor of adenosylhomocysteinase,[79] is readily phosphorylated by adenosine kinase and deaminated by ADA[80] so that treated cells receive little exposure to the nucleoside analog and its cytotoxicity is primarily due to its conversion to the nucleotides.

The carbocyclic analog (38) of 3-deazaadenosine is not deaminated nor phosphorylated in detectable amounts.[34] It is converted, in at least one cell type, but not detectably in others, to the carbocyclic analog of 3-deazaadenosylhomocysteine (64, X = CH_2). Its K_i for adenosylhomocysteinase is slightly lower than that of 3-deazaadenosine. Despite the fact that carbocyclic 3-deazaadenosine does not seem to be phosphorylated, it may be as cytotoxic to H.Ep. #2 cells in culture as carbocyclic adenosine (Table 11), even though it is highly active against vaccinia virus, as well as VSV and a number of other RNA viruses at non-toxic levels.[81]

62

63

64

Table 11.   Cytotoxicity of 3- and 7-Deazaadenine and Some Derivatives Thereof[a]

	IC_{50}, μM		
	H.Ep. #2	L1210	L1210 (AK$^-$)
3-Deazaadenine	7		
3-Deazaadenosine	20[b]		
Carbocyclic 3-deazaadenosine	< 1[b]	4	1
7-Deazaadenine	>70		
7-Deazaadenosine	0.002		
Carbocyclic-7-deazaadenosine	2	4	> 40
Carbocyclic adenosine	0.7		20[c]

[a]See Footnotes to Table 7.   [b]Unchanged in the presence of 2'-deoxycoformycin.   [c]H.Ep. #2/MeMPR.

The carbocyclic analog (42) of tubercidin (7-deazaadenosine) is slightly less cytotoxic to H.Ep. #2 cells than the 3-deaza compound but equally toxic to L1210 cells, except that L1210/MeMPR cells deficient in adenosine kinase are highly resistant to the 7-deaza compound but not to the 3-deaza compound. These results indicate that nucleotide formation is not involved in the action of the 3-deaza compound but that the 7-deaza compound must be converted to its 5'-phosphate. Since neither of these compounds are true nucleosides, they cannot be phosphorylized to the bases. Carbocyclic 7-deazaadenosine has no antiviral activity and is a poor inhibitor of adenosylhomocysteinase relative to the 3-deaza compound.

III. SUMMARY

The haloadenine nucleosides, with one exception, are highly resistant to deamination by adenosine deaminase. Some of the adenosine analogs are substrates for adenosine kinase and are cytotoxic but show no selectivity for cancer cells, although definite data is not yet available for 9 (X = Cl, F). The 5'-substituted ribonucleosides are substrates for methylthioadenosine phosphorylase and show selective cytotoxicity for cells containing this enzyme. The 2'-deoxyribonucleosides and the arabino-nucleosides are phosphorylated by 2'-deoxycytidine kinase. Their triphosphates selectively inhibit DNA synthesis and are curative of murine leukemias.

The carbocyclic analogs of 3- and 7-deazaadenosine are also resistant to deamination and are cytotoxic, even though only the 7-deaza compound appears to be phosphorylated. The 3-deaza compound has been found to be converted, to a small extent, to the analog of adenosylhomocysteine in a single cell line, and is a potent antiviral agent against vaccinia and several RNA viruses.

ACKNOWLEDGMENTS

I wish to acknowledge the contributions of Drs. L. L. Bennett, Jr., Dr. F. M. Schabel, Jr., and Dr. W. C. Coburn, Jr., and their staffs. I would also like to acknowledge our collaborations with Drs. D. A. Carson, M.-C. Huang, R. L. Blakley, T. M. Savarese, and R. E. Parks, Jr. Most of all, I am indebted to Ms. A. T. Shortnacy, S. D. Clayton, and Dr. J. A. Secrist III, who are responsible for producing these interesting nucleosides.

REFERENCES

1. Davoll, J., and Lowy, B. A., J. Am. Chem. Soc. 74:1563 (1952).

2. Sato, T. Simadate, T., and Ishido, Y., Nippon Kagaku Zasshi. 81:1440, 1442 (1960).

3. Montgomery, J. A., and Hewson, K., J. Heterocycl. Chem. 1:213 (1964).

4. Robins, M. J., and Uznanski, B., Can. J. Chem. 59:2601 (1981).

5. Schaeffer, H. J., and Thomas, H. J., J. Am. Chem. Soc. 80:3738 (1958).

6. Montgomery, J. A., and Hewson, K., J. Med. Chem. 12:498 (1969).

7. Christensen, L. F., Broom, A. D., Robins, M. J., and Bloch, A., J. Med. Chem. 15:735 (1972).

8. Montgomery, J. A., Shortnacy, A. T., Arnett, G., and Shannon, W. M., J. Med. Chem. 20:401 (1977).

9. Glaudemans, C. P. J., and Fletcher, Jr., H. G., J. Am. Chem. Soc. 87:4636 (1965).

10. Keller, F., Botvinick, I. J., and Bunker, J. E., J. Org. Chem. 32:1644 (1967).

11. Trummlitz, G., Repke, D. B., and Moffatt, J. G., J. Org. Chem. 40:3352 (1975).

12. Montgomery, J. A., Clayton, S. D., and Shortnacy, A. T., J. Heterocycl. Chem. 16:157 (1979).

13. Holmes, R. E., and Robins, R. K., J. Am. Chem. Soc. 86:1242 (1964).

14. Ikehara, M., and Kaneko, M., Tetrahedron. 26:4251 (1970).

15. Brentnall, H. J., and Hutchinson, D. W., Tetrahedron Lett. 2595 (1972).

16. Ikehara, M., Ogiso, Y., and Maruyama, T., Chem. Pharm. Bull. 25:575 (1977).

17. Reist, E. J., Calkins, D. F., Fisher, L. V., and Goodman, L., J. Org. Chem. 33:1600 (1968).

18. Ryu, E. K., and MacCoss, M., J. Org. Chem. 46:2819 (1981).

19. Montgomery, J. A., and Hewson, K., J. Am. Chem. Soc. 79:4559 (1957); idem, ibid., 82:463 (1960).

20. Bitterli, P., and Erlenmeyer, H., Helv. Chim. Acta. 34:835 (1951).

21. Ikehara, M., and Yamada, S., Chem. Pharm. Bull. 19:104 (1971).

22. Kobayashi, Y., Kumadaki, I., Ohsawa, A., and Murakami, S., J. Chem. Soc., Chem. Commun. 430 (1976).

23. Olah, G. A., Welch, J. T., Vankar, Y. D., Nojima, M., Kerekes, I., and Olah, J. A., J. Org. Chem. 44:3872 (1979).

24. Robins, M. J., and Uznanski, B., Can. J. Chem. 59:2608 (1981).

25. Ikehara, M., and Fukui, T., Biochim. Biophys. Acta, 338:512 (1974).

26. Lessor, R. A., and Leonard, N. J., J. Org. Chem. 46:4300 (1981).

27. Robins, M. J., and Wilson, J. S., J. Am. Chem. Soc. 103:932 (1981).

28. Pankiewicz, K., Matsuda, A., and Watanabe, K. A., J. Org. Chem. 47:485 (1982).

29. Carson, D. A., Wasson, D. B., Kaye, J., Ullman, B., Martin, Jr., D. W., Robins, R. K., and Montgomery, J. A., Proc. Natl. Acad. Sci. USA. 77:6865 (1980).

30. Huang, M.-C., Hatfield, K., Roetker, A. W., Montgomery, J. A., and Blakley, R. L., Biochem. Pharmacol. 30:2663 (1981).

31. Krenitsky, T. A., personal communication.

32. Krenitsky, T. A., Koszalka, G. W., Tuttle, J. V., Rideout, J. L., and Elion, G. B., Carbohydr. Res. 97:139 (1981).

33. Huang, M.-C., Montgomery, J. A., Thorpe, M. C., Stewart, E. L., Secrist III, J. A., and Blakley, R. L., submitted.

34. Montgomery, J. A., Clayton, S. J., Thomas, H. J., Shannon, W. M., Arnett, G., Bodner, A. J., Kim, I.-K., Cantoni, G. L., and Chiang, P. K., J. Med. Chem. 25:626 (1982).

35. Montgomery, J. A., and Hewson, K., J. Med. Chem. 10:665 (1967).

36. Shealy, Y. F., and Clayton, J. D., J. Am. Chem. Soc. 91:3075 (1969).

37. Cermak, R. C., and Vince, R., Tetrahedron Lett. 22:2331 (1981).

38. Chilson, O. P., and Fisher, J. R., Arch. Biochem. Biophys. 102:77 (1963).

39. Simon, L. N., Bauer, R. J., Tolman, R. L., and Robins, R. K., Biochemistry. 9:573 (1970).

40. Maguire, M. H., and Sim, M. K., Eur. J. Biochem. 23:22 (1971).

41. Agarwal, R. P., Sagar, S. M., and Parks, Jr., R. E., Biochem. Pharmacol. 24:693 (1975).

42. Bennett, Jr., L. L., Vail, M. H., Chumley, S., and Montgomery, J. A., Biochem. Pharmacol. 15:1719 (1966).

43. Bennett, Jr., L. L., Schnebli, H. P., Vail, M. H., Allan, P. W., and Montgomery, J. A., Mol. Pharmacol. 2:432 (1966).

44. Schnebli, H. P., Hill, D. L., and Bennett, Jr., L. L., J. Biol. Chem. 242:1997 (1967).

45. Miller, R. L., Adamczyk, D. L., Miller, W. H., Koszalka, G. W., Rideout, J. L., Beacham III, L. M., Chao, E. Y., Haggerty, J. J., Krenitsky, T. A., and Elion, G. B., J. Biol. Chem. 254:2346 (1979).

46. Shigeura, H. T., Boxer, G. E., Sampson, S. D., and Meloni, M. L., Arch. Biochem. Biophys. 111:713 (1965).

47. Zimmerman, T. P., Rideout, J. L., Wolberg, G., Duncan, G. S., and Elion, G. B., J. Biol. Chem. 251:6757 (1976).

48. Zimmerman, T. P., Wolberg, G., Duncan, G. S., Rideout, J. L., Beacham III, L. M., Krenitsky, T. A., and Elion, G. B., Biochem. Pharmacol. 27:1731 (1978).

49. Zimmerman, T. P., Biochem. Pharmacol. 28:2533 (1979).

50. Zimmerman, T. P., Wolberg, G., Duncan, G. S., and Elion, G. B., Biochemistry. 19:2252 (1980).

51. Zimmerman, T. P., Deeprose, R. D., Wolberg, G., and Duncan, G. S., Biochem. Biophys. Res. Commun. 91:997 (1979).

52. Montgomery, J. A., and Hewson, K., J. Med. Chem. 13:427 (1970).

53. LePage, G. A., Worth, L. S., and Kimball, A. P., Cancer Res. 36:1481 (1976).

54. Johns, D. G., and Adamson, R. H., Biochem. Pharmacol. 25:1441 (1976).

55. Adamson, R. H., Zaharevitz, D. W., and Johns, D. G., Pharmacology. 15:84 (1977).

56. Glazer, R. I., Cancer Chemother. Pharmacol. 4:227 (1980).

57. Plunkett, W., Benjamin, R. S., Keating, M. J., and Freireich, E. J., Cancer Res. 42:2092 (1982).

58. Brockman, R. W., Schabel, Jr., F. M., and Montgomery, J. A., Biochem. Pharmacol. 26:2193 (1977).

59. Brockman, R. W., Cheng, Y.-C., Schabel, Jr., F. M., and Montgomery, J. A., Cancer Res. 40:3610 (1980).

60. Skipper, H. E., Schabel, Jr., F. M., Mellett, L. B., Montgomery, J. A., Wilkoff, L. J., Lloyd, H. H., and Brockman, R. W., Cancer Chemother. Rept. 54:431 (1970).

61. Tseng, W.-C., Derse, D., Cheng, Y.-C., Brockman, R. W., and Bennett, Jr., L. L., Molec. Pharmacol. 21:474 (1982).

62. Dow, L. W., Bell, D. E., Poulakos, L., and Fridland, A., Cancer Res. 40:1405 (1980).

63. Plunkett, W., Chubb, S., Alexander, L., and Montgomery, J. A., Cancer Res. 40:2349 (1980).

64. El Dareer, S. M., Struck, R. F., Tillery, K. F., Rose, L. M., Brockman, R. W., Montgomery, J. A., and Hill, D. L., Drug Metab. and Dispo. 8:60 (1980).

65. White, E. L., Shaddix, S. C., Brockman, R. W., and Bennett, Jr., L. L., Cancer Res. 42:2260 (1982).

66. Krenitsky, T. A., Tuttle, J. V., Koszalka, G. W., Chen, I. S., Beacham III, L. M., Rideout, J. L., and Elion, G. B., J. Biol. Chem. 251:4055 (1976).

67. Bennett, Jr., L. L., and Allan, P. W., Cancer Res. 36:3917 (1976).

68. Montgomery, J. A., and Hewson, K., J. Heterocycl. Chem. 9:445 (1972).

69. Montgomery, J. A., Shortnacy, A. T., and Thomas, H. J., J. Med. Chem. 17:1197 (1974).

70. Parks, Jr., R. E., Stoeckler, J. D., Cambor, C., Savarese, T. M., Crabtree, G. W., and Chu, S. H. in "Molecular Actions and Targets for Cancer Chemotherapeutic Agents". (A. C. Sartorelli, J. S. Lazo, and J. R. Bertino, eds.), p. 229. Academic Press, New York, 1981.

71. Backlund, Jr., P. S., and Smith, R. A., J. Biol. Chem. 256:1533 (1981).

72. Backlund, Jr., P. S., Chang, C. P., and Smith, R. A., J. Biol. Chem. 257:4196 (1982).

73. Rousseau, R. J., Townsend, L. B., and Robins, R. K., Biochemistry. 5:756 (1966).

74. Montgomery, J. A., Shortnacy, A. T., and Clayton, S. D., J. Heterocycl. Chem. 14:195 (1977).

75. Chiang, P. K., Richards, H. H., and Cantoni, G. L., Mol. Pharmacol. 13:939 (1977).

76. Bader, J. P., Brown, N. R., Chiang, P. K., and Cantoni, G. L., Virology. 89:494 (1978).

77. Bodner, A. J., Cantoni, G. L., and Chiang, P. K., Biochem. Biophys. Res. Commun. 98:476 (1981).

78. Shealy, Y. F., and Clayton, J. D., J. Am. Chem. Soc. 91:3075 (1969).

79. Guranowski, A., Montgomery, J. A., Cantoni, G. L., and Chiang, P. K., Biochemistry. 20:110 (1981).

80. Bennett, Jr., L. L., Allan, P. W., and Hill, D. L., Mol. Pharmacol. 4:208 (1968).

81. DeClercq, E., personal communication.

EXPERIMENTAL AND CLINICAL STUDIES ON 2'-FLUOROARABINOSYL PYRIMIDINES AND PURINE-LIKE C-NUCLEOSIDES[1]

J.H. Burchenal

B. Leyland-Jones

B. Watanabe

R. Klein

C. Lopez

J.J. Fox

Memorial Sloan-Kettering Cancer Center

New York, New York

[1]This work was supported the American Cancer Society Grant CH-27W; National Cancer Institute Grants CA-05826, CA-08748, CA-16534, CA-18601, CA-18856, and CA-27569; and the Hearst Found.

The pyrimidine nucleoside 1-β-D-arabinofuranosylcytosine

(Ara-C) has been the mainstay for many years in the treatment

of acute non-lymphocytic leukemia (ANLL), the most common acute

leukemia of adults. Given along with one of the anthracyclines,

daunorubicin or doxorubicin, with or without thioguanine, it is

considered first line treatment for this disease in almost all

protocols (1). It has certain drawbacks, however, in that it

is rapidly deaminated to the inactive Arabinosyluracil (Ara-U)

and for the best results must be given by continuous intravenous

administration over 5 to 7 days. In addition, the leukemic cells

eventually develop resistance to Ara-C and cease to respond. In

an attempt to avoid the rapid deamination two masked precur-

sors, 2,2'-anhydro-aracytosine (cyclocytidine) synthesized by

Hoshi, et al. (2) and 2,2'-anhydro-ara-5-fluorocytosine (AAFC),

synthesized by Fox, et al. (3) were tested in mouse leukemias

and found to be active i.p. or p.o. against Ara-C sensitive but

not against Ara-C resistant lines. Unlike Ara-C they were rela-

tively schedule independent. In patients, they were active by

single or twice daily injections in ANLL, but were ineffective

in patients whose leukemic cells had become resistant to Ara-C

(4). 5-Azacytidine, a close analog of cytidine, was synthesized

by the Prague group under Sorm (5) and was active in mouse leu-

kemias both sensitive to and resistant to Ara-C. In patients,

Karon, et al. (6) demonstrated this activity against Ara-C

resistant acute non-lymphocytic leukemia in both children and adults. Its limiting toxicity clinically was excessive nausea and vomiting. Hoping to avoid this toxicity on the assumption that it might have been due to the metabolic byproducts of the drug, Fox, et al (7) synthesized the more stable C nucleoside, pseudoisocytidine, (ψICR) an isostere of cytidine and of 5-azacytidine. ψICR also, is active against Ara-C resistant leukemias in mice. Unfortunately, in clinical trial ψICR produced moderate liver toxicity at doses which were not thought to be adequate for therapeutic effects and the study was temporarily discontinued although it had not been studied in patients with Ara-C resistant ANLL (8). It is of interest that laboratory studies (Table 1) in various lines of mouse leukemia have shown potentiation of relatively inactive small doses of 25-50 mg/kg of pseudoisocytidine by equally weak doses, 2.5 to 5.0 mg/kg, of 5-azacytidine (9) and since their clinical toxicities are quite dissimilar it is likely that the two might be given effectively in combination to patients at levels below the toxic doses of either compound but with a summation of the antileukemic effect.

Table 1. Potentiation of Antileukemic Effects of Pseudoiso-
          cytidine (ψICR) and 5-Oazacytidine (AzaCR) in
          Leukemia L1210

Compound	Dose: mg/kg q4dx3	Survival Time (days)	Average Δ weight day 10 (g/mouse)	ILS
control	--	9.9	3.0	--
ψICR	150	14.6	0.0	47
ψICR	100	12.7	+0.9	28
ψICR	50	11.7	+4.7	18
AzaCR	10	16.3	-1.5	64
AzaCR	5	17.1	-0.1	72
AzaCR	2.5	12.3	+3.7	24
ψICR + AzaCR	50 5	23.2	0.7	134
ψICR + AzaCR	50 2.5	24.5	+0.4	147
ψICR + AzaCR	25 2.5	15.0	+1.5	51

In the uracil series examples of other nucleosides which
have demonstrated clinical usefulness have been 5-iodo-
2'deoxy-uridine (IUDR) and trifluorothymidine as antiviral
agents in herpes keratitis and 5-fluoro-2'-deoxyuridine
(FUDR) by continuous intraarterial infusion in tumors
limited to the liver as demonstrated by Ensminger, et al.
(10). It is of interest in passing that it had been shown

in 1960 by Sullivan, et al. (11) that 5-fluorouracil by
continuous intravenous administration decreased in activity
whereas in contrast FUDR on continuous administration
increased tremendously in activity. This increase in
potency is being taken advantage of in the intra-arterial
infusion of liver metastases. Other nucleosides such as
arabino-syladenine (12) and particularly acycloguanosine
(Acyclovir)(13) have shown high activity against Herpes
type viruses by the local and parental route.

In recent years, Fox, Watanabe, et al. (14-16)
synthesized arabinosyl-cytosine and -uracil nucleosides,
substituted in 2' "up" (arabino) position and the 5 position
which have antileukemic and antiviral activity. Figure 1
shows the various substitutions in the 5 position made on
2'-fluoro Ara-C or Ara-U. The most interesting of these
turned out to be the 2'-fluoro-5-iodo Ara-C (FIAC) or 2'-
fluoro-5-methyl Ara-C (FMAC) and their corresponding Ara-U
derivatives (FIAU, FMAU). The cytosine derivatives were
inactive in strains of mouse leukemia resistant to Ara-C as
shown in Table 2 whereas the uracil derivatives were
relatively active. Generally speaking, these compounds
were much more active against cell lines of human leukemias
and Burkitt's tumors than they were against mouse leukemias.

Figure 1

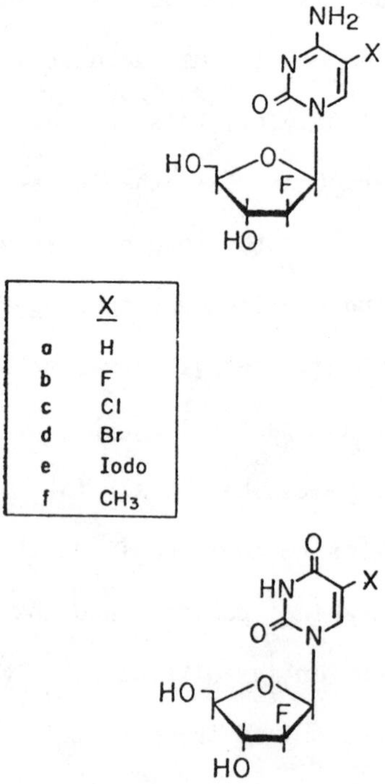

Table 2.   Comparison of in vitro Growth Inhibitory Effect of
           5-Iodo and 5-Methyl Substituted 2'F-ARA-C and
           2'F-ARA-U

ID50 in µG/ML

	P815	P815/Ara-C	L5178Y	L5178Y/Ara-C	L1210
			MOUSE LINES		
FIAC	10	>100	12	>100	5
FIAU	0.8	10	0.9	1.2	4
FMAC	0.5	>100	1.5	>100	0.25
FMAU	2.5	7	3.5	4.3	4

It is of interest that the antileukemic effect of the two cytosine derivatives could be blocked by deoxycytidine in mouse leukemia where relatively high concentrations of the antimetabolite were required, whereas they could not be blocked in the Burkitt's tumor cells where much lower levels of FIAC were needed.

In mouse leukemias, FIAC was without activity at 1,000 mg/kg daily but FMAU at 800 to 1,000 mg/kg by mouth or 800 mg/kg by ip were highly effective against P815/Ara-C resistant line and also effective but somewhat less so against L1210/ Ara-C (17). Not shown in Figure 2, FMAU at doses of 400 and 800 mg/kg every second day for six doses or pseudoisocytidine in the same schedule at 25 or 50 mg/kg were highly effective.

Figure 2

POTENTIATION OF THE ANTILEUKEMIC EFFECT OF FMAU BY ↓ICR IN LEUKEMIA P815/ARA-C
Doses in mg/kg q2dx5

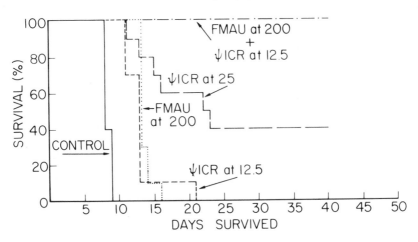

Alone FMAU at 200 mg/kg or pseudoisocytidine at 12.5 mg/kg
had relatively little effect against P815/Ara C. The combination of the two, however, showed marked potentiation within
this particular experiment with all animals living the 50
day period. This again suggests a potentially active clinical
combination for ψICR and FMAU.

In the purine type nucleosides area, Tubercidin and
Nebularine are two highly toxic nucleosides. In an attempt
to develop better compounds, Klein, et al. (18) have synthesized the C-nucleosides 9-deaza and 7,9-dideaza-7-thienoadeno-
sine. These compounds are highly active against various
mouse and human leukemias in tissue culture and are some of
the most toxic compounds that we have every worked with
(Table 3). In the mouse, however, there is very little spread
between the toxic and therapeutic activity. For this reason
other derivatives have been studied. Generally, the inosine
derivatives have been relatively inactive both in vitro and
in vivo whereas the 6-thioinosine analogs have been intermediate in toxicity and have demonstrated some chemotherapeutic effect in vivo. Based on the work of Paterson, et
al. (19) we have attempted to block the toxic effect of the
two adenosine derivatives by nitrobenzylthioinosine in the
mouse, but so far have not found the ideal dose which will
allow therapeutic without toxic effects. Further work is
required on this study.

Table 3. Comparison _in vitro_ Activity (ID$_{50}$'s in μg/ml) of Pyrrolo- and Thieno[3,2-d]pyrimidine C-Nucleosides in Mouse and Human Cell Lines

			P-815	L-1210	All-Raji	CCRF-CEM	HL-60
	Y=OH , R=β-D-Rib		>10(0%)	>100	>100	>100	>100
	Y=SH , R= "		5.6	-	-	-	-
X=NH	Y=SMe, R= "		0.3	-	-	-	-
	Y=NH$_2$, R= "		0.001	0.0008	0.002	0.0008	0.0003
	Y=OH , R=β-D-Rib		2.5	3.3	4.2	0.9	0.4
	Y=SH , R= "		0.5	0.6	0.9	1.6	0.4
	Y=SMe, R= "		0.03	0.07	0.03	0.006	0.008
X=S	Y=SMe, R=α-D-Rib		>10(0%)	-	-	-	-
	Y=SMe, R=H		"	-	-	-	-
	Y=NH$_2$, R=β-D-Rib		0.0003	0.0006	0.002	0.0005	0.0005
	Y=NH$_2$, R=α-D-Rib		0.6*	-	-	-	-
	Y=NH$_2$, R=H		>10(0%)	-	-	-	-

*Presence of some β-isomer is possible.

To return to the 2'-fluoro pyrimidines, Fox, Watanabe and Lopez (14-16) have reported previously on the antiherpes activity of several of these 2'5-disubstituted Ara-C and Ara-U (Figure 3).

J. H. Burchenal *et al.*

Figure 3

CAPACITY OF 2'-FLUORO-2'-DEOXYARABINOSYLCYTOSINE AND -URACILS
TO SUPPRESS HSV-1 REPLICATION IN MONOLAYERS OF VERO CELLS

X	ANTIVIRAL ACT.[A] IN μG/ML					CYTOTOXICITY, ID50 IN μG/ML[B]	
	0.01	0.1	1.0	10	100	L5178Y	P815
			CYTOSINE NUCLEOSIDES				
H	–	+	+	+++	++++	0.05	0.05
F	–	+	++	+++	ND[C]	0.5	0.4
CL	–	–	–	+++	ND	1.4	1.0
BR	–	+	++	+++++	ND	10	> 10
I	+	++	++++	+++++	+++++	15	10
CH3	–	–	+	++	++	1.5	.0.8
			URACIL NUCLEOSIDES				
H	–	–	++	++	ND	> 10	> 10
F	–	++	++++	+++++	ND	1.0	0.7
CL	–	–	–	++++	ND	1.4	3.4
BR	–	–	++	++++	ND	0.9	1.6
I	–	+	+++	++++	ND	0.9	0.8
CH3	+	++	++	+++++	+++++	1.6	1.9

[A] PERCENT REDUCTION OF HSV-1 TITER: >90% = +; >99% = ++; >99.9% = +++;
>99.99% = ++++; COMPLETE OBSERVATION OF HSV-1 REPLICATION = +++++.
(–) = < 90% REDUCTION OF HSV-1 TITER.
[B] CONCENTRATION REQUIRED FOR 50% INHIBITION OF GROWTH OF CELLS IN VITRO AT
37° C FOR 96 HOURS.
[C] ND = NOT DONE

WATANABE, K., REICHMAN, U., HIROTA, K., LOPEZ, C. AND FOX, J. J. MED. CHEM.
22(1):21-24, 1979.

In vitro, FIAC was the most potent giving a 90% reduction of
HSV-1 and 2 in Vero cell monolayers at levels as low as 0.01
μg/ml. The 5-bromo analog of FIAC and also FMAU, FIAU and
FFAU were almost as potent as FIAC in vitro with relatively
minimal cytotoxicity against uninfected Vero cells. Watanabe,
et al. (17) had noted that the 2'-fluoro substituent in the

"up" (arabino) configuration in the sugar moiety of these arabinonucleosides confers better antiherpes activity than does 2'-hydroxyl or 2'-hydrogen substituent. For example, 2'-fluoro-5-iodo cytidine (the ribo-isomer of FIAC) with the 2'-fluoro in the "down" configuration was at least 1,000 times less effective in inhibition of HSV-1 replication. It appears again from all these studies that the nature of the substitution on the C2' and C5 position plays an important role in determining antiviral and cytotoxic activity.

The studies of Lopez, et al. (15,16) with FIAC indicate that its mechanism of antiviral activity in HSV is dependent in large measure on the viral specified thymidine kinase. In their studies FIAC was about 6,000 fold more active against the wild type HSV-1 than against the mutant strain lacking this viral enzyme. In contrast, the cytotoxicity of FIAC in normal cells seems to depend on a cellular deoxycytidine kinase since this can be reversed by deoxycytidine and not by thymidine. On the other hand, the antiviral activity is only reversed by high concentration of thymidine but not by deoxycytidine which suggests that in its antiviral activity FIAC is acting as an analog of thymidine and not deoxycytidine. A summary of Lopez studies on the mechanism of action of FIAC are given in Table 4.

TABLE 4.   Conclusions

---

1. FIAC is preferentially phosphorylated by the HSV-1
   specified dThd kinase while, in uninfected cells,
   FIAC is probably activated by a dCyd kinase.

2. FIAC does not inhibit strongly phosphorylation of
   dThd and thus is probably not active by inhibiting
   viral dThd kinase. (Active at some point after
   this).

3. Selective inhibition of {^{3}H}dThd incorporation into
   HSV-1 DNA as compared to cellular DNA suggests possible
   selective inhibition of viral DNA polymerase. (May
   also be incorporated into viral DNA).

4.   FIAC strongly inhibits {^{3}H}dUrd incorporation into
     DNA in HSV-1 infected but not in uninfected Vero
     cells. By contrast, {^{3}H}dThy incorporation in both
     Vero and HSV-1 infected Vero cells was weakly
     inhibited. These data suggest that thymidylate
     synthetase may be inhibited by FIAC.

FIAC was also active against cytomegalovirus (CMV) in WI-38
fibroblast monolayers, although CMV has no virus-specified
t.k. Watanabe, et al. (20) have reported on the antiviral
activity of a group of 2'-deoxy-2'-halogeno Ara-C (type) and
Ara-U derivatives substituted at the C-5 of the pyrimidine
with bromine, iodine or methyl (Table 5). Their data also
show that in general the fluoro function in the 2' "up"

(arabino) configuration offers better antiviral activity than does the corresponding 2'-chloro or 2'-bromo analogs. It is noteworthy that 2'-chloro-5-methyl and 2'-chloro-5-iodo-Ara-C showed much more activity against HSV-2 than against HSV-1. No such differences were seen in the Ara-U series. This suggests that some of these 2'Cl Ara-C analogs might serve as useful probes to distinguish between HSV types 1 and 2 in the differential diagnosis of HSV infections in man.

Table 5.   Antiherpetic Activity of Some 2'-Halogeno-5-Substituted Ara-Cytosine Nucleosides.

			ED_{50}* (µM)		ID_{50} (µM)
			HSV-1	HSV-2	
NUCLEOSIDE	X	R	(STRAIN 2931)	(STRAIN G)	
	F	H	0.12	0.3	0.6
	CL	H	0.11	3.8	0.006
	BR	H	3.2	4.4	0.2
	F	BR	0.19	0.02	5.0
	CL	BR	>100	>100	>100
	BR	BR	>100	>100	4.5
	F	I [FIAC]	0.01	0.01	8.6
	CL	I	3.4	0.09	3.5
	BR	I	9.5	2.2	0.2
	F	ME [FMAC]	0.64	0.8	0.8
	CL	ME	>100	0.22	20.0
	BR	ME	>100	26.2	34.0

* ED_{50} indicates effective dose to suppress viral replication by 50% (plaque reduction assay).

⊥ ID_{50} indicates the concentration necessary for 50% inhibition of growth of normal human lymphocytic cells.

FIAC also protected mice inoculated with HSV-1 and 2 (15,16). For this reason, clinical studies were initiated.

Phase I and II trials of FIAC in patients with herpes zoster
have been completed and at a dose of 200 mg/M^2 twice daily,
a marked beneficial effect consisting of rapid drying of
lesions and almost no appearance of new lesions after the
first 72 hours was noticed (21). In a double blind study
against arabinosyladenine by Leyland-Jones, et al. (22)
FIAC was significantly better in the rapidity of drying of
lesions, in the almost absence of new lesions after 72 hours
of therapy and the diminution in pain.

O'Reilly, et al. are now beginning a Phase II trial of
FIAC as prophylaxis of viral infections in the immunosuppressed
patients with bone marrow transplant, particularly against
CMV interstitial pneumonia.

Figure 4

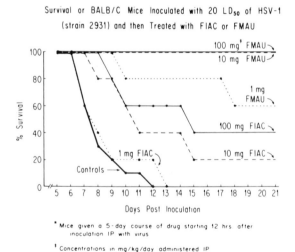

Survival or BALB/C Mice Inoculated with 20 LD$_{50}$ of HSV-1
(strain 2931) and then Treated with FIAC or FMAU

* Mice given a 5-day course of drug starting 12 hrs. after
inoculation IP with virus

† Concentrations in mg/kg/day administered IP

The original studies by Lopez, et al. (15,16) showed that although they were comparable in their activity in vitro, in vivo FMAU (1 mg/kg qd x 5) was approximately 100 fold more potent than FIAC (100 mg/kg qd x 5) (Figure 4). Price, et al. (20) using intraocular inoculation of 2.6 x 10 plaque forming units of the F strain of HSV 1, gave ^{14}C labeled FMAU on day 5 or 6 after the virus when brain invasion had already developed and sacrified 6 or 24 hours later. There was selective uptake and concentration of FMAU by infected cells allowing quantitative definition and mapping of HSV-1 infected structures in autoradiograms of brain sections. For this reason, it was decided by our Investigational New Drugs Committee that this compound also deserved clinical trial. More recently the data of Schinazi, Peters and Nahmias (24) have shown that FMAU at as little as 0.1 mg/kg (0.3 mg/M^2) twice daily for 4 days can protect mice inoculated intracerebrally with 67 LD of HSV-2. This is considerably lower than the level of FIAC (5 mg/kg), Ara-A (15 mg/kg) or Acylovir (30 mg/kg) which provided a similiar level of protection. Accordingly, Philips and his group (25) in our pharmacology laboratory, studied FMAU for pre-clinical toxicology in mice, rats and dogs. They found to their surprise that it was much more toxic in dogs than in mice or rats with a maximum tolerated dose of about 1.25 mg/kg (25 mg/M^2) daily for 10 days. For that reason,

clinical studies were started cautiously at extremely low
levels of 2 mg/M^2 daily for five days by intravenous push.
This dose has been gradually escalated to a tolerated level
of 32 mg/M^2. No antitumor effects have been noted at this
level but it has not been given as yet to patients with
Ara-C resistant acute leukemia (ANLL) (26). We now hope to
investigate continuous administration versus i.v. push and
also the activity of oral dosage. Blood levels will be monitored by HPLC and antiviral titrations. As mentioned above,
the most recent studies by Schinazi, et al. (24) show that
FMAU even at extremely low doses is highly effective even
against intracerebrally inoculated HSV-2. This suggests
that even though a therapeutic effect against Ara-C resistant
leukemias is not reached in patients, we may be well above
(perhaps even 1 to 2 logs above) the dose necessary for
antiviral effect. Since no myelosuppression has been noted
at the highest levels studied in man or in dogs, FMAU at
low levels might be ideal in the prophylaxis of interstitial
pneumonia in marrow transplant patients.

In summary, we feel that the pyrimidine and purine
nucleo-sides continue to offer a potentially rewarding field
for exploration for compounds with antitumor and antiviral
activity.

REFERENCES

1. Burchenal, J.H. Leukemia Overview in "Cancer: Achieve-
ments, Challenges, and Prospects for the 1980's" (J.H.
Burchenal and H.F. Oettgen, eds.), p. 249. Grune and
Stratton, New York, 1981.

2. Hoshi, A., Kanzawa, F., Kuretani, K., Saneyoshi, M.,
and Arai, Y.,Gann. 62:145 (1971).

3. Fox, J.J., Falco, E.A., Wempen, I., Pomeroy, D., Dowling,
M.D., and Burchenal, J.H., Cancer Res. 32:2269 (1972).

4. Burchenal, J.H., Kalaher, K., Clarkson, B., Kemeny,
N., Young, C., Fox, J., and Krakoff, I., in "Current
Chemotherapy" (W. Siegenthaler and L. Ruede, eds.),
p. 1206. American Society of Microbiology, Washington,
1978.

5. Piskala, A., and Sorm, F., Collection Czech. Chem.
Communl. 29:2060 (1964).

6. Karon, M., Sieger, K., Leimbrock, S., Finklestein, J.,
Nesbit, M., and Swaney, J., Blood 42:359 (1973).

7. Burchenal, J.H., Ciovacco, K.R., Kalaher, K.M., O'Toole,
T., Kiefner, R., Dowling, M.D., Jr., Chu, C., Watanabe,
K., Wempen, I., and Fox, J.J., Cancer Res. 36:1520
(1976).

8. Woodcock, T., Burchenal, J.H., Young, C.W., Chou, T.,
Philips, F.S., Wollner, N., Sternberg, S.S., and Tan,
C., Proc. Amer. Assoc. Cancer Res. 19:197, abstract
#788, (1977).

9. Burchenal, J.H., Ciovacco, K.R., Kalaher, K.M., O'Toole, T., Kiefner, F., Dowling, M.D., Jr., Chu, C., Watanabe, K., Wempen, I., and Fox, J.J., Proc. Amer. Assoc. Cancer Res. 16:143, abstract #569 (1976).

10. Keller, J., Ensminger, W., Niederhuber, J., Dakhil, S., Thrall, J., and Wheller, R., Proc. Amer. Soc. Clin. Oncol. 22:354 (1981).

11. Sullivan, R.D., Young, C.W., Miller, E., Glatstein, N., Clarkson, B., and Burchenal, J.H., Cancer Chemo. Reports 8:77 (1960).

12. Chien, L.T., Cannon, N.J., Charamella, L.J., Dismukes, W.D., Whitley, R.J., Buchanan, R.A., and Alford, C.A., Jr., J. Inf. Dis. 128:658 (1973).

13. Elion, G.B., Furman, P.A., Fyfe, J.A., de Miranda, P., Beauchamp, L., and Schaeffer, H.J., Proc. National Academy of Sciences (USA) 74:5716 (1977).

14. Watanabe, K.A., Reichman, W., Hirota, K., Lopez, C., Fox, J.J., J. Med. Chem. 22:21 (1979).

15. Lopez, C., Watanabe, K.A., Fox, J.J., Antimicrob. Agents Chemother. 17:803 (1980).

16. Fox, J.J., Lopez, C., Watanabe, K.A. in "Medicinal Chemistry Advances" (F.G. De Las Heras, ed.), p. 27. Pergamon Press, New York, 1981.

17. Burchenal, J.H., Chou, T-C., Lokys, L., Smith, R.S., Watanabe, K.S., Su, T-L., and Fox, J.J., Cancer Res. 42:2598 (1982).

18. Lin, M-I. and Klein, R.S., Tetrahedron Letters 22:25 (1981).

19. Lynch, T.P., Jakobs, E.S., Paran, J.H., and Paterson, A.R.P., Cancer Res. 41:3200 (1981).

20. Fox, J.J., Watanabe, K.A., Lopez, C., Philips, F.S., and Leyland-Jones, B. in "Proceedings of International Symposium on Herpes Viruses", Excerpta Medica, Amsterdam (In Press).

21. Young, C., Jones, B., Schneider, R., Armstrong, D., Tan, C., Lopez, C., Watanabe, K., Fox, J., and Philips, F., Proc. Amer. Assoc. Cancer Res. 22:165, abstract #656 (1981).

22. Leyland-Jones, B., Donnelly, H., Myskowski, P., Donner, A., Groshen, S., Clarkson, B., Lopez, C., Philips, F., Young, C., Armstrong, D., and Fox, J., (In Press).

23. Sato, Y., Price, R.W., et al., Science 217:1151 (1982).

24. Schinazi, R.F., Peters, J., and Nahmias, A.J., Antimicrob. Agents Chemother. (In Press).

25. Philips, F.S., Feinberg, A., Chu, T-C., Vidal, P.M., Su, T-L., Watanabe, K., and Fox, J.J. (In Press).

26. Leyland-Jones, B. Personal Communication.

2',5'-Oligoadenylates: Their Role in Interferon Action and Their
Potential as Chemotherapeutic Agents

Paul F. Torrence[1], Krystyna Lesiak[1], Jiro Imai[1],
Margaret I. Johnston[1] and Hiroaki Sawai[2]

[1] Laboratory of Chemistry, National Institute of Arthritis
Diabetes, and Digestive and Kidney Diseases
National Institutes of Health, Bethesda, Maryland

[2] Faculty of Pharmaceutical Sciences, University of Tokyo
Bunkyo-ku, Tokyo, Japan

When vertebrate cells are treated with interferon, a wide
variety of biological responses can occur. For instance,
interferon may establish an antiviral state in the treated cell[1],
prevent replication of intracellular parasites(e.g.,Rickettsiae)[2],
inhibit production of induced cellular enzymes (e. g., tyrosine
aminotransferase[3]), enhance production of induced cellular enzymes
(e.g., aryl hydrocarbon hydroxylase[4]), exhibit antitumor[5] or
antiproliferative[6] activity, prime cells for interferon production[7], increase the toxicity of double-stranded RNA[8], alter
cellular membranes[9], inhibit thymidine uptake[10], modulate the
immune system (e. g., enhancement of natural killer cell activity[11]), regulate differentiation[12,13], stimulate prostaglandin
formation[14] block the the mitogenic effect of epidermal growth
factor[15], or alter cell membrane phospholipid synthesis.[16]

Interferons have been cloned and expressed in prokaryotic
cells[17-27], purified to homogeneity[28-36], crystallized[37] and their
amino acid sequence determined [30-36] by both classical methodology
and cDNA sequences[17-27]. Sufficient leukocyte A interferon has
been made available from recombinant DNA technology so that the
first clinical trials with this material have been reported[38].
The availability of pure molecular species of interferons is
obviously a great boon to interferon research. The recent advances
have, however, added an additional dimension of complexity since
it is now realized that there are more than dozen human leukocyte
(α)interferons as well as at least one human fibroblast(β)inter-
feron and at least one immune (γ)interferon[20,24,39-41]. Moreover,
different interferons may vary considerably in their biological
activities; for instance, the ratio of antiviral activity to
antiproliferative activity is remarkably different among various
α(leucocyte)interferons[42].

✗   The interferon cellular defense mechanism as it operates in
nature involves two distinct stages[1,43]. Infection of a cell by
virus (or stimulation by any of a variety of apparently dissimilar
substances[44]) may trigger the gene for interferon production.
Interferon is then excreted from the infected cell and binds
to another potential host cell thereby eliciting what is probably a membrane-mediated signal[45] for activation of one or more
genes responsible for establishing the antiviral state.

How does interferon affect the replication of viruses? The
answer to this question seems to depend upon the nature of the
virus examined (Table 1). What is clear from inspection of Table
1 is that interferons may have more than one mode of action but
that beyond any doubt, translation is a primary target of interferon action in the case of a number of different viruses.

Once translation as site of interferon action had been
discovered, attention turned to investigation of cell-free protein
synthesis systems to establish the exact mechanism of such
translational inhibition[43,46-50]. In the content of this current

Table 1. Site of Inhibition of Virus Replication By Interferon

Virus and classification	Effect of interferon treatment of host cell
Sendliki forest (positive strand RNA)	Parental RNA not translated.
Vesicular Stomatitis Virus (negative Strand RNA)	Reduction of all VSV proteins; Non-infectious virus particles may be produced as a result of reduced viral glycoproteins.
Reovirus (dsRNA)	Accumulation of reovirus specific mRNA's, dsRNA's and proteins inhibited. Reovirus mRNA methylation blocked.
Vaccinia Virus ((dsDNA)	Second stage uncoating and DNA synthesis inhibited, viral RNA synthesis stimulated; all caused by primary effect on viral protein synthesis. Cycloheximide imitates the effect of interferon in this system.
SV40 (DNA tumor virus)	Treatment with interferon before injection gives blockade of virus RNA synthesis; later treatment gives no RNA synthesis inhibition, only inhibition of translation. Also SV40 mRNA is overmethylated.
Murine sarcoma or leukemia virus (RNA tumor viruses) chronically infected cells	Effect on late stages of virus maturation; amount of virus budding from cell membrane reduced; non-infectious particles formed.

Sources used in compiling this table included references 43, 45-47.

discussion, one of the most provocative findings came from the laboratory of Ian Kerr[51]. Double-stranded RNA (dsRNA) was found to be a potent inhibitor of translation in extracts of interferon-treated cells but had no significant effect on protein synthesis in extracts of cell which had not been interferon-treated. This result was of even greater interest in view of earlier work which established that under certain conditions, while treatment of cells with interferon alone did not block translation in the resulting cell extracts, interferon treatment followed by virus infection could evoke a translational blockade in the resulting extracts. This data suggested that an intermediate of virus replication was able to trigger an antiviral state which was latent after interferon treatment. These latter observations and those of Kerr's group could be reconciled since dsRNA is generaged during the replication of some RNA viruses and possibly DNA viruses also. Thus, the theory could be formulated that interferon generated a latent antiviral state which could be activated by the generation of dsRNA formed during the virus's life cycle or introduced by the virus itself (e. g., reovirus)[46-50].

How might dsRNA act to inhibit protein synthesis? At least two distinct mechanisms exist to account for translational inhibition caused by dsRNA. The first of these pathways (Figure 1) is termed the 2-5A system and owes its origins to the original investigations of Kerr and colleagues who found that a low molecular weight inhibitor of protein synthesis was generated when extracts of interferon-treated mouse cell were incubated with dsRNA and ATP[53,54]. Kerr and Brown[55] were able to determine the structures of this low molecular weight inhibitor as a series of unique 2'5'-linked oligoadenylates of the general formula $ppp5'A2'(p5'A)n$ ($1 \leq n \leq 15$) (Figure 2), also called 2-5A. 2-5A itself is an extremely potent inhibitor of translation, causing a 50% reduction in protein synthesis in encephalomyocarditis-virus RNA programmed L cell extracts at a concentration of $10^{-9}\underline{M}$ in oligomer. The enzyme, 2-5A synthetase, which synthesizes 2-5A, is activated by dsRNA

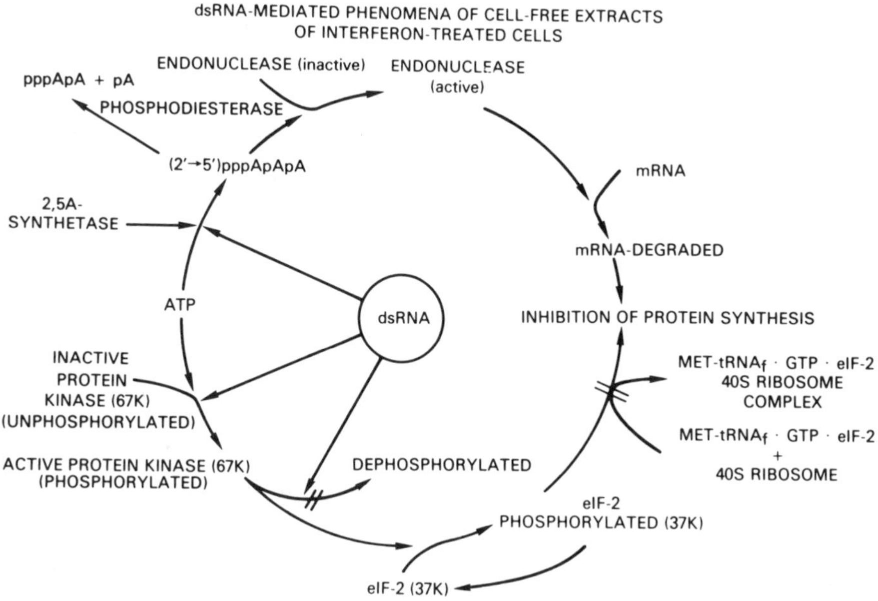

dsRNA-MEDIATED PHENOMENA OF CELL-FREE EXTRACTS
OF INTERFERON-TREATED CELLS

pppApA + pA

ENDONUCLEASE (inactive)　ENDONUCLEASE (active)

PHOSPHODIESTERASE

(2′→5′)pppApApA

mRNA

2,5A-SYNTHETASE

mRNA-DEGRADED

ATP　dsRNA

INHIBITION OF PROTEIN SYNTHESIS

INACTIVE PROTEIN KINASE (67K) (UNPHOSPHORYLATED)

MET-tRNA$_f$ · GTP · eIF-2
40S RIBOSOME COMPLEX

MET-tRNA$_f$ · GTP · eIF-2
+
40S RIBOSOME

ACTIVE PROTEIN KINASE (67K) (PHOSPHORYLATED)

DEPHOSPHORYLATED

eIF-2 PHOSPHORYLATED (37K)

eIF-2 (37K)

Figure 1. Mechanisms involved in the inhibition of translation caused by dsRNA in extracts of interferon-treated cells.

in a structure-dependent manner[56,57] and has been purified to homogeneity[58,59].

The following theory has developed [46,48-50] regarding the role of 2-5A in the mechanism of antiviral action of interferons. When a cell is treated with interferon, elevated levels of 2-5A synthetase are induced. Then dsRNA, formed as an intermediate or byproduct of virus reproduction, activates the synthetase to generate 2-5A from ATP. The 2-5A thus formed activates a latent cellular endoribonuclease which then degrades mRNA, thereby inhibiting translation. This 2-5A-dependent endonuclease is not

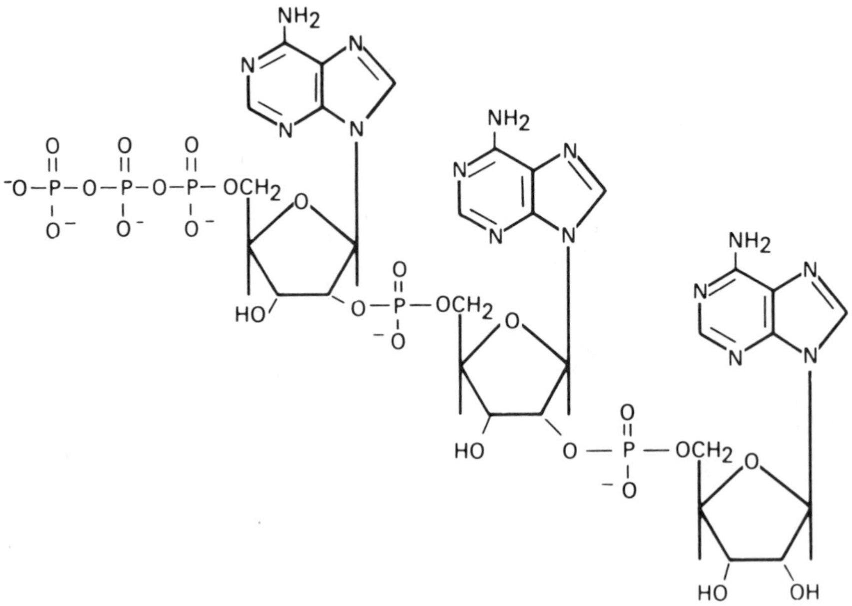

5'-O-TRIPHOSPHORYLADENYLYL-(2'→5')-ADENYLYL-(2'→5')-ADENOSINE

"2-5 A"

Figure 2. Structure of 2-5A

very specific since it can cleave single-stranded RNA at the
3'side of UU, UG and UA sequences[60,61]. Support for this concept
of the mechanism of interferon action comes from:

    i)   the observation that 2-5A, introduced into
        the cell by artificial means, can halt virus
        replication[62-64];

  ii)   the detection of 2-5A, in amounts sufficient
        to block protein synthesis, in interferon-
        treated cells infected with encephalomycarditis
        virus or reovirus[65,66];

iii)   ribosomal RNA cleavage, peculiar to 2-5A-
       dependent nuclease action, occuring in
       interferon-treated cells injected with
       virus[67].

Both dsRNA and 2-5A might be considered alarmones since they
represent signal molecules which are associated with the cellular
stress of virus infection.

A second and independent mechanism that may be involved in
the mechanism of the antiviral activity of interferons is the
protein kinase pathway[46,48-50]. (Figure 1). A central feature
in this scheme is also a dsRNA-dependent enzyme, a kinase (protein
P_1 kinase) which phosphorylates protein P_1 (67-69K daltons in
mouse cells). According to this scenario, the protein P_1 kinase
also is induced to higher levels in the cell by treatment with
interferon. The dsRNA from virus infection then activates this
kinase which subsequently phosphorylates protein P_1 and the small
α-subunit of eukaryotic initiation factor 2 (eIF-2)[53,68-73].
This phosphorylation may be regulated by a phosphoprotein phosphatase which may also be dsRNA dependent[74,75]. Phosphorylation
of the α-subunit of eIF-2 decrease its ability to engage in
initiation complex formation with a 40 S ribosomal subunit, GTP
and met-tRNA$_f$[76]. While the exact basis of the phenomenon is not
yet established, one possibility is that phosphorylation of eIF-2
may prevent its normal interaction with recycling factor which
usually operates to displace tightly bound GDP from a catalytically inactive complex released upon 60 S ribosomal subunit
joining to the 48 S preinitiation complex[76,77].

Both the protein P_1 kinase pathway and 2-5A pathway can be
demonstrated in extracts of interferon-treated cells, and the
question arises which, if either, of the two mechanisms are
responsible for the inhibition of translation caused by dsRNA in
extracts of interferon-treated cells and therefore, by
inference in interferon-treated and virus-infected cells? One
possible approach would be to employ an inhibitor of one or the

other pathway; this would permit evaluation of importance of the
other enzyme system in mediating the effects of dsRNA. Indeed,
inhibitors of the 2-5A synthetase[78] and protein P_1 kinase[78,79]
have been described, but they are not selective enough to be of
use in this situation. Since the triphosphate, ppp5'A2'p5'A2'-
p5'A, and the diphosphate, pp5'A2'p5'A2'p5'A, are equally effective as activators of the 2-5A-dependent endonuclease, and
since the monophosphate p5'A2'p5'A2'p5'A is devoid of such
activity, we asked if the failure of p5'A2'p5'A2'p5'A to activate
the endonuclease may be due to its inability to bind to the
endoribonuclease or because it could bind to, but not activate the
endonuclease.[80] To address this question experimentally, we
determined whether or not p5'A2'p5'A2'p5'A could prevent the translational inhibitory activity of ppp5'A2'p5'A2'p5'A. In fact,
p5'A2'p5'A2'p5'A was an effective antagonist of 2-5A action
(Figure 3) since it could bind to the endonuclease, and thereby
prevent the binding of 2-5A itself.[81] Importantly, the
5'-monophosphate, p5'A2'p5'A2'p5'A did not affect the extent of
dsRNA-mediated phosphorylation of endogenous eIF-2α, protein P_1 or
added rabbit reticulocyte eIF-2α, nor did it affect the site of
phosphorylation of eIF-2α in cell-free extracts of interferon-
treated mouse L cells. Under identical conditions, however,
p5'A2'p5'A2'p5'A was able to prevent the protein synthesis inhibitory effects of poly(I)·poly(C), thereby showing that 2-5A is the
primary mediator of the translational inhibitory effects of
poly(I)·poly(C) in cell-free extracts of interferon-treated mouse
L cells (Figure 4). Even more effective reversal of the action
of poly(I)·poly(C) has been obtained with other analogues of
p5'A2'p5'A2'p5'A[82] (vide infra). Furthermore, the protein
synthesis inhibitory action of other dsRNA's such as poly(A)·
poly(U) or _Penicillian chrysogenum_ dsRNA could be prevented with
p5'A2'p5'A2'p5'A or its analogues (P. F. Torrence and J. Imai,
unpublished observations).

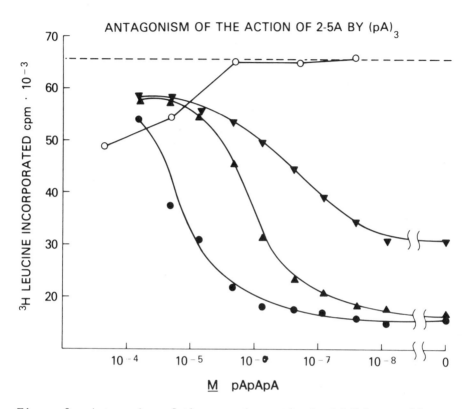

Figure 3. Antagonism of the protein synthesis inhibitory effects of 2-5A by p5'A2'p5'A2'p5'A. The concentration of 2-5A was held constant at 1.7 n$\underline{M}$ (▼), 17n$\underline{M}$ (▲) or 170 n$\underline{M}$ (●), while the concentration of p5'A2'p5'A2'p5'A was varied from 200 µ$\underline{M}$ to 8n$\underline{M}$. Open circles (0) represent the effect of adding various concentrations of p5'A2'p5'A2'p5'A to the reaction mixture. The dashed line (---) indicates the level of translation obtained in the absence of any additions.

Added support for the idea that 2-5A may be the primary mediator of the protein synthesis inhibition caused by dsRNA has come from studies[83] with a synthetic nucleic acid, polyadenylic acid·poly(2'-fluoro-2'-deoxyuridylic acid), poly(A)·poly(dUfl).[84] This polynucleotide, unlike others so far examined, activates only

the protein P_1 kinase but not 2-5A synthetase, making it a valuable tool for evaluation of the relative importance of these
pathways.

Addition of poly(A)·poly(dUf1) to rabbit reticulocyte
extracts results in a profound inhibition of globin synthesis in a
manner similar to other dsRNA's such as P. chrysogenum, poly(A)·
poly(U) or poly(I)·poly(C). Poly(A)·poly(dUf1) induced phosphorylation of eIF-2α in the same manner as hemin deprivation or other

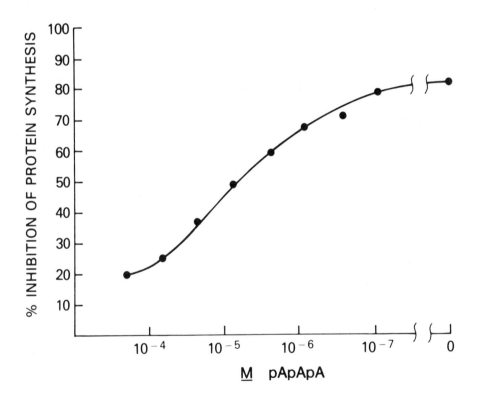

Figure 4. Antagonism by p5'A2'p5'A2'p5'A of the protein synthesis
inhibitory effects of poly(I)·poly(C) in extracts of interferon
treated mouse L cells. The concentration of poly(I)·poly(C) was
held constant at $2 \cdot 10^{-6}$Mp and the concentration of trimer varied.
TCA insoluble radioactivity was determined after 90 min of translation.

more common dsRNA's. It did not, however, lead to generation of
2-5A; moreover the lysates used in these experiments were nearly
insensitive to exogenous 2-5A. These data argued that the inhibition of protein synthesis caused by poly(A)·poly(dUfl) in rabbit
reticulocyte lysates is due to the phosphorylation of eIF-2α and
is, therefore, similar in mechanism to the inhibition caused by
hemin deprivation or nucleic acids such as P. chrysogenum dsRNA.

The behavior of poly(A)·poly(dUfl) in extracts of interferon-
treated mouse L cells was radically different from its behavior in
reticulocyte lysates since it was absolutely inactive as an inhibitor of translation. Surprisingly, however, poly(A)·poly(dUfl)
stimulated phosphorylation of eIF-2α as well as protein P_1 to the
same extent as did poly(I)·poly(C). In addition, peptide mapping
of ^{32}P-labelled endogenous or exogenous eIF-2α showed that the
site(s) of phosphorylation were the same in the case of both
poly(I)·poly(C) and poly(A)·poly(dUfl). It can therefore be
concluded that activation of the protein P_1 kinase to lead to
phosphorylation of protein P_1 and eIF-2α and/or protein P_1 may
play at best a minor role in mediating the inhibition of protein
synthesis caused by double-stranded RNA, and, to the extent such
cell-free systems reflect the in vivo situation, such phosphorylation may not be critical in the mechanism of the antiviral state
induced by interferon.

The conclusion that it may be the 2-5A system and not the
kinase pathway that operates to establish the interferon-induced
antiviral state is supported by several bits of circumstantial
evidence. Although this study has been challenged recently,[85]
Epstein et al[86] determined that interferon-treated NIH-3T3 cells
(clone 1) were not protected by interferon against encephalomyocarditis virus infection and that such cell extracts showed
efficient phosphorylation of eIF-2α and protein P_1 but were
insensitive to protein synthesis inhibition by poly(I)·poly(C)
and were deficient in 2-5A-activated endonuclease activity.
Interferon or oxidized glutathione treatment of mouse L cells

induced the dsRNA dependent kinase and subsequent eIF-2α
phosphorylation; however, only interferon induced an antiviral
state.[87]   In human amnion cells, Samuel and Knutson[88] reported
that interferon induced the kinase and blocked reproduction
of vesicular stomatitis virus but not replication of reovirus.   When
intact interferon-treated mouse L cells were treated with dsRNA,
protein P_1 phosphorylation could be detected, but when such cells
were infected with mengovirus or vesicular stomatitis virus, no
protein P_1 phosphorylation could be seen.[89]

At this point, however, a cautionary note needs to be added.
Indeed, not all experiments are perfectly consistent with the role
of the 2-5A system in the antiviral action of interferon.   For
instance, some cell lines, such as the human HEC-1[90] and the mouse
K_{balb},[91] have very high levels of 2-5A synthetase and apparently
normal levels of 2-5A-activated endonuclease, but are not in an
antiviral state.   In addition, the HEC-1 line does not become
resistant to virus even after interferon treatment.[90]   These kinds
of anomalies are, of course, difficult to reconcile with the
purported role of the 2-5A system; nonetheless it should be borne
in mind that the true test would include a determination of the
in vivo functioning of the 2-5A system, i.e., whether 2-5A is
indeed synthesized and whether it activates nuclease to destroy
RNA.   Moreover, it must be noted, and this might be intuitively
obvious upon consideration of the multitudinous biological effects
of interferon (vide ante), that interferon may inhibit virus
replication by means of more than one mechanism and that the
mechanism which finally obtains may be determined by the cell
as well as the invading virus.[48]

Some circumstantial evidence has suggested that, in addition
to its role in the antiviral action of interferon, 2-5A may play
some role in cell regulation and/or differentiation.   High levels
of 2-5A synthetase activity have been described in cell types as
diverse as lymphocytes,[92] lymphoblastoid cells,[95] reticulocytes,[94]

cells from dog[95] or mouse liver,[96] or estrogen-stimulated and
withdrawn chick oviduct.[97] In addition, there is evidence that
the 2-5A synthetase may be inducible by agents other than interferon such as by dexamethasone in lymphoblastoid cells,[98] by
cortisteroid in cultured chick embryo tendon fibroblasts,[99] by
dimethylsulfoxide or sodium butyrate in maloney sarcoma virus-
transformed murine BALB/c cells[100] or by dimethyl sulfoxide in
cultures of Friend erythroleukemia cells.[101] In the last
instance, exogenous interferon was produced and could be directly
responsible for the increased synthetase levels.[101] Finally,
Revel and colleagues have suggested that 2-5A is at least partly
involved in the inhibition of cell proliferation caused by
interferon.[102]

The potent translational inhibitory activity of 2-5A when
studied in cell-free systems or when introduced into cells by
permeabilization, transfection or microinjection techniques, the
antiviral activity exhibited upon introduction into cells by the
latter techniques, its likely role in the antiviral and antiproliferative actions of interferon, and its possible involvement in
cellular regulation and/or differentiation suggest that it would
be extremely valuable to be able to employ 2-5A or some derivative
thereof in intact cells. This would provide increased opportuni-
ties to pursue the biologic roles of the 2-5A system as well as
permit investigation of a potentially novel approach to virus or
cancer chemotherapy. In addition, such studies need not be
confined to 2-5A itself since antagonists of 2-5A, such as p5'A2'-
p5'A2'p5'A, also would be useful in establishing the biological
role of the 2-5A system and may also find a role in the treatment
of what has been termed interferon-induced disease.[101-106]

Realization of the above goals is severely limited by two
established considerations. The first is that the 2-5A molecule
has a relatively brief half-life in biological systems since
it is rapidly degraded by a 2',5'-phosphodiesterase which

degrades 2-5A from its 2'-end, giving 5'AMP and 5'ATP as products.
Studies on the partially purified phosphodiesterase have shown it
to have a slight preference for cleavage of 2',5'-phosphodiester
bonds as opposed to 3',5'-bonds but not to be dependent on base
sequence.[109] This same enzyme can cleave the CCA terminus of tRNA
thereby reducing its amino acid acceptance capacity.[109] A second
equally important consideration is that the 2-5A molecule itself,
because of its ionic character, cannot penetrate the eukaryotic
cell. Chemical modification of the 2-5A molecule provides an
approach to solution of these problems, but the nature of such
alterations must be compatible with binding to and/or activation
of the 2-5A-dependent endonuclease.

Interaction of various analogues of 2-5A with the 2-5A-
dependent endonuclease has been followed using three different
assay methods. i) The first involves determination of the
ability of the modified oligonucleotide 5'-monophosphate to
antagonize or prevent the translational inhibitory effects of
2-5A in a mouse L cell-free protein synthesis system programmed
with encephalomyocarditis virus RNA.[80] Since this antagonism is
the result of the competition of the oligonucleotide analogue for
the 2-5A binding site of the endonuclease,[81] it provides a measure
of the ability of the analogue to bind to the endonuclease. In
such experiments, the usual protocol was to choose a constant
concentration (20 n$\underline{M}$) of 2-5A as the minimal concentration needed
to affect maximum inhibition of translation. Then, increasing
quantities of modified oligomer were added to the reaction
mixtures to determine the concentration of analogue required to
bring about a 50% reversal of the inhibition of protein synthesis
caused by 2-5A. This latter value was used to compare the relative oligonucleotides as antagonists of 2-5A action. ii) The
second assay was the determination of the ability of the
corresponding isomeric oligoadenylate 5'-triphosphate to
inhibit protein synthesis in the same encephalomyocarditis virus

Table 2. Antagonism of 2-5A Action by Oligoadenylates

Compd	Oligomer	Concentration required to effect 50% reversal of 2-5A action
"core" oligomers		
1	A2'p5'A	$> 1 \cdot 10^{-3}$ $\underline{M}$
2	A2'p5'A2'p5'A	$1 \cdot 10^{-4}$ $\underline{M}$
3	A2'p5'A2'p5'A2'p5'A	$1 \cdot 10^{-5}$ $\underline{M}$
4	A2'p5'A2'p5'A2'p5'A2'p5'A	$7 \cdot 10^{-6}$ $\underline{M}$
5'-monophosphorylated oligomers		
5	p5'A2'p5'A	$> 2 \cdot 10^{-4}$ $\underline{M}$
6	p5'A2'p5'A2'p5'A	$1 \cdot 10^{-6}$ $\underline{M}$
7	p5'A2'p5'A2'p5'A2'p5'A	$1 \cdot 10^{-6}$ $\underline{M}$
8	p5'A2'p5'A2'p5'A2'p5'A2'p5'A	$5 \cdot 10^{-7}$ $\underline{M}$
3',5'-bisphosphorylated oligomers		
9	p5'A2'p5'A2'(3')p	$5 \cdot 10^{-6}$ $\underline{M}$
10	p5'A2'p5'A2'p5'A2'(3')p	$3 \cdot 10^{-7}$ $\underline{M}$

RNA-directed L cell-free system used above. This approach simply involved addition of increasing amounts of analogue to the cell-free system and subsequent determination of the concentration of oligomer required to effect a half-maximal inhibition of protein synthesis. iii) The third method relied upon the assay developed in Ian Kerr's laboratory[81] and uses ppp5'A2'p5'A2'p5'A2'p5' A3'-[^{32}P]p5'C3'p as a radioactive probe which, in nitrocellulose binding assay, can be displaced by oligonucleotides that bind to the 2-5A site on the endonuclease. In this case, the concentration of oligomer needed to prevent 50% of the radiolabeled probe from binding to the endonuclease-nitrocellulose complex was used as a measure of affinity to the 2-5A-dependent endonuclease. All of the compounds described herein have been evaluated by at least two of the above procedures.

Table 2 presents the results of a study of the influence of oligoadenylate chain length and extent of phosphorylation on antagonistic activity.[82] Inspection of Table 2 reveals that a 5'-terminal monophosphate moiety is a critical albeit not essential requirement for antagonistic activity. No unphosphorylated oligomer (compds. 1-4) is as active as the 5'phosphorylated oligoadenylates (compds. 5-8) and removal of the 5'-monophosphate moiety leads to a 10-100-fold increase in the concentration needed to prevent 2-5A action; nevertheless, although trimer core (compd. 8) is 100x less active than p5'A2'p5'A2'p5'A, tetramer and pentamer cores (compds. 3 and 4) are only 10x less active than tetramer or pentamer 5'-phosphates. In view of the fact that elongation of the oligonucleotide chain beyond three nucleotide residues does not lead to large increases in antagonistic capacity, it seems likely that the reduced activity of tetramer or pentamer core might be due to the possibility that the first (from the 5'-end) internucleotide phosphate bond may "slip" onto the binding site normally occupied by the 5'-monophosphate group.

The data of Table 2 also show that at least three adenosine
nucleotide residues appear to be required for maximal activity as
antagonist of 2-5A action and thus for binding to the 2-5A-
activated endonuclease. The relative lack of activity of compd.
5 compared to 6 is witness to this point. In addition, the
activities of the "core" oligomers lend support to such a concept
if the "slippage" mechanism outlined above holds. In this case,
reduced, albeit significant activity is seen with the tetramer
core (compd. 3) since it provides the necessary additional nucleotide of p5'A2'p5'A2'p5'A. It is likely, however, that only a
portion of the 2'-terminal AMP residue of p5'A2'p5'A2'p5'A is
needed for activity. Removal of adenosine from the tetramer
1 to give a 2'(3')-terminally phosphorylated trimer (compd 10)
leads to an increase in biolgical activity (probably due to
increased phosphodiesterase resistance[110]). A similar removal
of an adenosine residue from the trimer 6 gave a 2'(3')-terminally
phosphorylated dimer (compd. 9) which was only 5x less active than
p5'A2'p5'A2'p5'A. Again, unequivocal interpretation of this
result is complicated by differences in phosphodiesterase
resistance, but it seems likely that significant endonuclease
binding may be possible even when only a phosphate group of the
terminal 2'-AMP residue of p5'A2'p5'A2'p5'A remains.

Also apparent from Table 2 is that beyond three nucleotides,
no substantial increase in antagonistic capacity (or endonuclease
binding) may occur. Thus adding adenosine nucleotide residues to
p5'A2'p5'A2'p5'A to give tetramer or pentamer monophosphates (7
and 8) has at most a marginal effect on activity.

The outstanding structural feature of the 2-5A molecule
is the presence of 2',5'-phosphodiester bonds: the existence of
such unique linkages in a molecule with powerful regulatory effects
infers their necessity. Nonetheless, the question can be posed:
How critical are the 2',5'-phosphodiester bonds of 2-5A for its
interaction with the 2-5A-activated endonuclease and its conse-
quent action as an inhibitor of translation? In addition, a

corollary question can be posed: would 3-5A, i.e., the 2-5A
analogue in which the phosphodiester bond are 3',5'-linked,
possess any biological activities similar to 2-5A? To answer
these questions, we synthesized and evaluated the biochemical
properties of various isomers of 2-5A in which one or more of the
2',5'-phosphodiester bonds have been replaced by 3',5' linkages.[111]

Two separate approaches were employed to prepare the
phosphodiester linkage isomers needed for this study. The first
method was that due to Sawai and collaborators[112] and consisted
of lead ion catalyzed polymerization of adenosine 5'-phosphoro-
imidazolide to give 5'-monophosphorylated trimers which were
converted to the corresponding 5'-triphosphates by conversion to
the phosphoroimidazolide followed by reaction with tri(n-butylammonium) pyrophosphate.[113,114] A second method involved T4
polynucleotide kinase[115] catalyzed phosphorylation of A3'p5'A3'-
p5'A to p5'A3'p5'A3'p5'A which was converted to the 5'-triphos-
phate in the same manner as described above. Assigned structures
were determined by enzymatic digestion patterns with such enzymes
as RNase P_1, RNase T_2, venom phosphodiesterase or bacterial
alkaline phosphatase, periodate oxidation and base elimination,
and ^{31}P and ^1H NMR.

Pertinent data regarding these synthetic linkage isomers can
be found in Table 3. Whether the results were judged in terms of
the ability of the isomeric oligonucleotide 5'-mono- or
-triphosphates to inhibit protein synthesis, to prevent the
translational inhibitory effects of 2-5A, or to prevent binding
of labeled probe to the endonuclease, it was clear that replacement of a single 2',5'-phosphodiester bond of 2-5A with a 3',5'-
phosphodiester bond leads to decreased ability to bind to or
activate the 2-5A-dependent endonuclease. This was most
dramatically demonstrated in the trimer series; thus, when a
single phosphodiester bond of ppp5'A2'p5'A2'p5'A (11) was replaced
by a 3',5'-linkage (i.e. 12, 13, 14 or 15), there resulted a 20-
50-fold decrease in activity. Within the limits of error of these

Table 3. Biological Activities of Phosphodiester Linkage Isomers of 2-5A

Compd	Oligomer	Antagonism of 2-5A[a] action by 5'-monophosphate	Inhibition of translation[a] by 5'-triphosphate	Competition of 5'triphosphate with radiolabeled probe in endonuclease binding assay[a]
11	ppp5'A2'p5'A2'p5'A	–	1	1
12	p5'A3'p5'A2'p5'A	20	–	–
13	ppp5'A3'p5'A2'p5'A	–	28	48
14	p5'A2'p5'A3'p5'A	24	–	–
15	ppp5'A2'p5'A3'p5'A	–	25	48
16	p5'A3'p5'A3'p5'A	>> 45[b]	–	–
17	ppp5'A3'p5'A3'p5'A	–	10^5	13,000

a. Data presented in terms of relative activity i.e., the relative amount needed to achieve a half-maximal effect. The higher the value, the less active was the oligomer.

b. No prevention of 2-5A action at highest concentration tested ($3 \cdot 10^{-5} \underline{M}$).

assays, it did not matter which of the two phosphodiester bonds
was replaced. However, when both 2',5'-phosphodiester bonds of
2-5A trimers were replaced by 3',5'-bonds there was nearly complete
loss of biological activity in the 3',5'-linked analogues (i.e,
16 and 17).

The preceding data show that the 2',5'-phosphodiester bonds
of the 2-5A molecule are critical for optimal binding to and
activation of the 2-5A-dependent endonuclease. Substitution of
either of the phosphodiester linkages of ppp5'A2'p5'A2'p5'A with
3',5'-bonds results in an order of magnitude or more loss of
binding or activation ability. Assuming that binding of one
phosphodiester bond site is independent from binding at the others,
then these data could be used to predict that replacement of both
2',5'-phosphodiester bonds of (pp)p5'A2'p5'A2'p5'A with 3',5'-
linkages should lead to a greater than two orders of magnitude
loss in binding or activation ability. The behavior of what can
be called 3-5A (compd 17) is in complete accord with prediction.

Other studies (vide infra and reference 116) on
cordycepin and per-O-methylated analogues of 2-5A have
indicated the importance of the 3'-hydroxyl group in binding
to and activation of the 2-5A-dependent endonuclease. Phosphodiester bond isomerization in an obvious dramatic disturbance of this important oligonucleotide binding/activation
site. In addition, oligonucleotide conformation also may play
some role in binding to or activation of the endonuclease.
Various physiochemical studies[117,119] have indicated that the
dinucleotides A2'p5'A and A3'p5'A differ in important conformational parameters. For instance, the geometry of the base-
stacked conformation of A2'p5'A at room temperature differs
from that of A3'p5'A. This could be related to the rather
unusual syn glycosidic torsion angle assumed by the first
(5'-terminal)nucleotide residue of A2'p5'A. In A3'p5'A, this
torsion angle is in the preferred anti range. Such a
conformational difference has been suggested as the reason why

2',5'-linked polymers cannot form stable helical nucleic acids.[119]

Recently, Doetsch et al[120] reported the preparation of a putative cordycepin analogue of 2-5A; specifically, ppp5'(3'dA)2'p5'(3'dA)2'p5'(3'dA). In lysed rabbit reticulocytes, they found this cordycepin 2-5A analogue was a more potent inhibitor of protein synthesis than was 2-5A trimer triphosphate, and in the same assay, they reported this new analogue to be equipotent with 2-5A tetramer triphosphates. It also was reported[121] that the "core" of this analogue, i.e., (3'dA)2'p5'(3'dA)2'p5'(3'dA), like the "core" of 2-5A itself, could prevent transformation of human lymphocytes infected by Epstein-Barr virus. On basis of such results, it was suggested that cordycepin "core" analogue could replace or supplement interferon treatment of cells, thereby implying a mechanism of action for the cordycepin analogue involving the 2-5A system. No independent confirmation of those findings have been forthcoming. Although Charubala and Pfleiderer[122] reported an unambigous chemical synthesis of the cordycepin 2-5A core, they did not convert it to the 5'-triphosphate, a step necessary to determine its ability to interact with the endonuclease. Thus, in collaboration with Dr. Hiroaki Sawai of the University of Tokyo, we undertook a study of the chemical synthesis and biological activity of the trimeric and tetrameric cordycepin analogues of 2-5A.[123]

Cordycepin analogues of 2-5A were prepared by dicyclohexyl carbodiimide-induced polymerization of cordycepin 5'-monophosphate (Figure 5) and the reaction products were separated on QAE-Sephadex with a triethylammonium bicarbonate gradient. A series of oligomers of the general formula p5'(3'dA)2'[p5'(3'dA)]$_n$ (n = 1-5) was obtained. Two of these, the trimer (n = 2) and tetramer (n = 3), were converted to the corresponding 5'-di and -triphosphate via the phosphoimidazolide method.[113,114] These mono- and di- and triphosphate

products were homogenous by HPLC, TLC and paper chromatography
Digestion of them with bacterial alkaline phosphatose and
snake venom phosphodiesterase gave cordycepin as the only
detectable product. Proton NMR of the monophosphate showed the
requisite number of characteristic adenine ring protons and
anomeric protons. ^{31}P NMR spectra of the cordycepin trimer and
tetramer triphosphates displayed the requisite number of
internucleotide phosphate resonances as well as the characteristic α, β and γ resonances of the triphosphate moiety.
Finally, we were fortunate to obtain samples of (3'dA)2'p5'-
(3'dA)2'p5'(3'dA) from Dr. Wolfgang Pfleiderer of the University of Konstanz and Dr. Robert Glazer of the National Cancer
Institute. Cordycepin "core" analogues from the latter source
had been prepared by Collaborative Research (Boston) by a
route similar to that used by Charubala and Pfleiderer.[122]
When our synthetic ppp5'(3'dA)2'p5'(3'dA)2'p5'(3'dA) was
digested with bacterial alkaline phosphotase, it gave a product
identical by HPLC and TLC to the samples of cordycepin "core".
Finally, material from Dr. Pfleiderer's laboratory was 5'-phos-
phorylated with T4 polynucleotide kinase to give a product
identical by HPLC with our synthetic p5'(3'A)2'p5'(3'dA)2'p5'-
(3'dA).

When the cordycepin trimer or tetramer di- or triphosphates
(p)pp5'(3'dA)2'p5'(3'dA)2'p5'(3'dA) or (p)pp5'(3'dA)2'p5'-
(3'dA)2'p5'(3'dA)p5'(3'dA), were evaluated for their ability
to inhibit protein synthesis in L cell extracts, none of
the analogues displayed significant inhibition of translation
at concentrations up to 10^{-5}M under conditions where 2-5A
itself blocked translation by 50% at 10^{-9}M (Figure 6).[123]
Since the original report of Doetsch et al[120] was based on
observations in reticulocyte lysates, the cordycepin trimer
and tetramer triphosphates also were checked in that system.

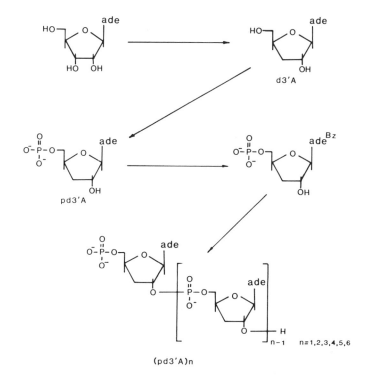

CHEMICAL SYNTHESIS OF CORDYCEPIN OLIGOMERS

Figure 5. Conversions involved in the chemical synthesis of a cordycepin analogue of 2-5A.

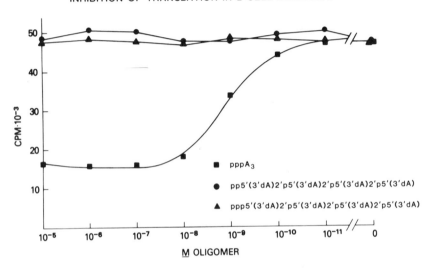

INHIBITION OF TRANSLATION IN L CELL EXTRACTS

Figure 6. Effect of cordecypin tetramer diphosphate (●) or triphosphate (▲) compared to effect of 2-5A (■) on translation in extracts of mouse L cells.

In such lysates (Figure 7),[123] ppp5'(3'dA)2'p5'(3'dA)2'p5'-(3'dA) was devoid of activity but the tetramer, ppp5'(3'dA)-2'p5'(3'dA)2'p5'(3'dA)2'p5'(3'dA), not reported by Doetsch et al[120], had at best 1/100 of the protein synthesis inhibition activity of 2-5A. These findings make it unlikely that cordycepin trimer triphosphate, ppp5'(3'dA)2'p5'(3'dA)2'p5'-(3'dA), is as effective as 2-5A as an inhibitor of translation. It is, therefore, also improbable that the reported ability of

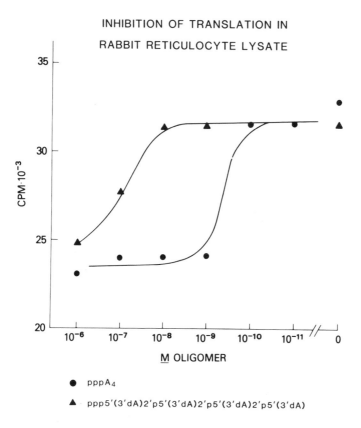

Figure 7. Comparison of the behaviors of 2-5A tetramer triphosphate (●) and the corresponding cordecypin analogue (▲) on
protein synthesis in rabbit reticulocyte lysates.

cordycepin trimer core to block transformation of human
lymphocytes after E-B virus infection is due to its conversion
to the 5'-triphosphate followed by activation of the 2-5A-
dependent endoribonuclease. Likewise, since such cordycepin
2-5A analogues cannot activate the endonuclease, there is
no theoretical basis to expect that they could replace
interferon in chemotherapeutic situations.[120,121,124]

Although such cordycepin analogues of 2-5A were not able
to induce translational inhibition, they were able to interact
with the 2-5A-activated endonuclease. This was the case for
the 5'-monophosphates, p5'(3'dA)2'p5'(3'dA)2'p5'(3'dA) and
p5'(3'dA)2'p5'(3'dA)2'p5'(3'd)2'p5'(3'dA), as well as the
corresponding 5'-di- and 5'-triphosphates. This was
established by the ability of such analogues to prevent the
translational inhibitory effects of 2-5A (e.g. Figure 8) and
to displace labeled probe from the endonuclease (e.g. Figure
9). Therefore, replacement of the 3'-hydroxyl moieties on all
the ribose residues of the 2-5A trimer or tetramer triphosphate resulted in an oligomer which could bind to the 2-5A-
activated endonuclease but could not activate it to degrade RNA.
This represents a distinct separation of the structural parameters
which govern binding to the endonuclease from the structural parameters involved in endonuclease activation. Such spearation of
binding and activation phases in endonuclease action has been
seen earlier but at the level of oligonucleotide phosphorylation;[80-82] i.e., unphosphorylated 2-5A core binds very
weakly to the endonuclease and is a very poor antagonist of
2-5A action, the 5'-monophosphate, p5'A2'p5'A2'p5'A, binds
well to the endonuclease and is an antagonist of 2-5A action,
and the triphosphate, 2-5A itself, binds strongly to the
endonuclease and is a potent endonuclease activator. The cordycepin analogues, however, provide the first instance in which
this discrimination between binding and activation can be
observed at the level of modification to the building block units
of the oligonucleotide. Lastly these data demonstrate that the
3'-hydroxyl groups of 2-5A are not critical in determining
binding to the endonuclease, but are critical determinants
of enzyme activation.

Modification to the 2-5A structure described thus far have
included the effects of chain length, phosphorylation state and
the nature of the ribose-phosphate backbone. We also have

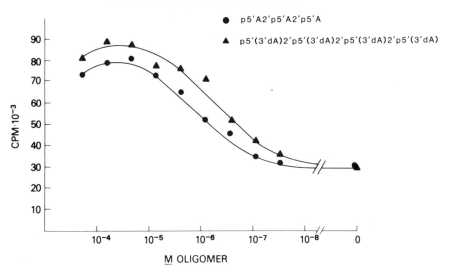

Figure 8. Prevention of 2-5A action: comparison of
p5'A2'p5'A2'p5'A (●) and p5'(3'dA)2'p5'(3'dA)2'p5'(3'dA)2'p5'-
(3'dA) (▲). Condition the same as Figure 6.

studied modification to the base moieties of 2-5A. These have
included alterations to or replacement of the adenine rings.
Analogues of 2-5A in which the adenine ring was replaced by
uracil, cytosine or hypoxanthine were prepared by lead ion
catalyzed polymerization in the laboratory of Dr. Sawai.[125,126]
The 8-bromo analogue of the 5-monophosphate of 2-5A core, i.e.,
p5(br[8]A)2'p5'(br[8]A)2'p5'(br[8]A), was also obtained by lead
ion catalyzed polymerization.[127] Finally, the fluorescent

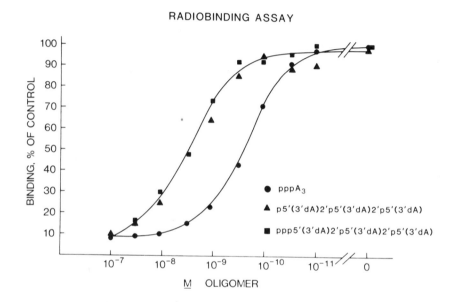

RADIOBINDING ASSAY

Figure 9. Ability of cordycepin trimer monophosphate (▲),
cordycepin trimer triphosphate (■) and 2-5A (●) to displace radiolabeled ppp5'A2'p5'A2'p5'A2'p5'A3'p5'C3'p from the 2-5A-activated
endonuclease of L cells.

etheno analogues of 2-5A and the 2-5A core monophosphate were
synthesized by reaction of p5'A2'p5'A2'p5'A with chloroacetal-
dehyde.[128]

As detailed in Table 4, replacement of the adenine rings
2-5A by pyrimidine rings (compds. 18 and 19) or alteration of
the adenine ring by substitution of a 6-hydroxyl for a 6-amino
moiety (compds. 20 and 21) or addition of a third ring at the
N_1 and N_6 portions in the etheno anlogues (compds. 22 and 23)
resulted in dramatic losses of biological activity whether

Table 4. Effect of Base Modifications on the Biological
Activity of 2-5A and 2-5A Core Monophosphate

Cmpd.	Oligomer	Antagonism of 2-5A action by monophosphate[a]	Inhibition of translation by 5'-triphosphate[a]	Competition of 5'-triphosphate with radiolabeled probe in endonuclease binding assay[a]
18	p5'U2'p5'U2'p5'U	100	-	2200
19	p5'C2'p5'C2'p5'C	15[b]	-	6200
20	p5'I2'p5'I2'p5'I	70	-	5900
21	ppp5'I2'p5'I2'p5'I	-	10,000	560
22	p5'(εA)2'p5'(εA)2'p5(εA)	100	-	-
23	ppp5'(εA)2'p5'(εA)2'p5'(εA)	-	10,000	5000

a. Data presented as relative activity; i.e., the relative amount needed to achieve a half-maximal effect. The greater the value, the less active was the oligomer.

b. Compd. 19 was inactive at the highest concentration that could be tested. This amounted to $2 \cdot 10^{-5}$ M since higher concentrations caused inhibition of translation.

defined as antagonistic capacity, translational inhibition or
endonuclease binding (Table 3). While in the aforementioned
analogues, all of the bases of 2-5A were replaced by a given
substitution, Drocourt et al[129] have found that compounds of
the general formula ppp5'A2'p5'A2'p5'N, where N may be any of
the common nucleosides, could not activate the endonuclease
but could antagonize 2-5A action. This latter result is
consistent with our previous observation that such oligomers as
p5'A2'p5'A2'p are capable of antagonism of 2-5A action.

From the foregoing considerations, it can be concluded
that certain regions of the 2-5A molecule are strategically
involved in binding to and/or activation of the 2-5A activated
endonuclease. Those regions include the 5'-phosphate or
triphosphate residue, the heterocyclic bases and the ribose-
phosphate backbone. It is clear, however, that the 2-terminus
will tolerate considerable changes without loss of ability
to bind to the endonuclease as judged by antagonism studies.
This area of the molecule might be expected to be subject to
chemical modifications which would not affect adversely the
biological activities of the resulting 2-5A analogue.

Our approach to modification of the 2-terminus of the 2-5A
molecule is patterned after chemistry first investigated by
Khym[130] and later by Brown and Read.[131] This involved
periodate oxidation, then Schiff base formation with hexylamine
and finally reduction with cyanoborohydride. This converts the
2'-terminal ribose moiety to a morpholine ring system. This
modification seemed attractive for two reasons: i) The
periodate oxidation/Schiff base formation/borohydride reduction
cycle had been applied successfully in the field of immuno-
chemistry to prepare nucleotide-protein conjugates to be used
as haptens. ii) Further modification at the 2'-terminus is
possible if amines are used that possess a functional group
compatible with the initial oxidation-reduction cycle.

Model studies were initiated with adenine and ATP to convert them to what we have termed hexylamine "tailed" [*] derivatives (Figure 10) by the periodate oxidation/Schiff base formation/borohydride reduction cycle. Compound <u>24</u> could be obtained in 90% yield by this procedure and its structure could be established by proton NMR. In addition to two singlets assigned to adenine ring protons at δ 7.99 and 8.09 ppm (Figure

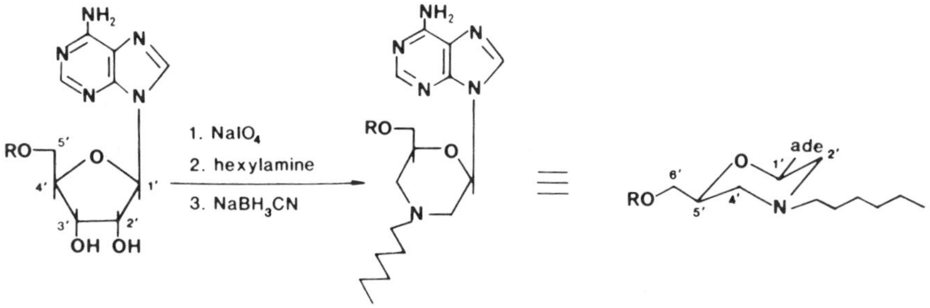

24. R= H

25. R = ppp –

26. R = A2'p –

27. R = p5'A2'p –

28. R = p5'A2'p5'A2'p –

29. R = p5'A2'p5'A2'p5'A2'p–

<u>Figure 10</u>. Scheme for the preparation of "tailed" 2-5A derivatives.

11) compound 25 showed characteristic aliphatic chain protons at δ 1.0 ppm. The side-chain methylene adjacent to the morpholine nitrogen appeared as a triplet at 2.40 ppm. The anomeric protein(H1') of the 3-azahexopyranose ring appeared at

---

[1]The hexylamine modification resembles a "tail" in molecular models of structural formulae.

δ 5.76 ppm as a doublet of doublets (J_1 = 3 Hz, J_2 = 10 Hz) and the 2'-methylene protons were found at δ 3.08 and 2.47 ppm as a doublet of doublets (J_1 = 3 Hz, J_2 = 10 Hz) and a triplet (J = 10 Hz). A highly split methyne proton (H5') appeared at δ 3.85 ppm and was coupled to H6' methylene protons at δ 3.55 ppm (doublet, J = 5 Hz) and H4' protons at 2.87 ppm (broad doublet, J = 11 Hz) and 2.06 ppm (triplet, J = 11 Hz). Taking into account the earlier work of Brown and Read,[131] and the result of elemental analysis ($C_{16}H_{26}N_6O_2$), the above data allowed the structure of 24 (Figure 10) structure to be assigned. Similar NMR patterns were obtained for the ATP derivative (25) but were not so clearly resolved due to downfield shifts caused by the presence of the triphosphate moiety; however, digestion of 25 with alkaline phosphatase gave 24 as the only product. For characterization of the preparation of 26 to 29, proton NMR spectra revealed the requisite number of purine and anomeric protons (Figure 11) and the oligomers were digested with a mixture of snake-venom phosphodiesterase and alkaline phosphatase to yield adenosine and compd 24 in the expected ratios.

Such "tailed" 2-5A core monophosphate analogues, compds 28 and 29, were significantly more active as antagonists of 2-5A action than was p5'A2'p5'A2'p5'A itself (Figure 12).[133] This antagonistic effect was manifested through the same mechanism as that of p5'A2'p5'A2'p5'A since 29 acted to prevent degradation of encephalomyocarditis virus RNA.[133] Furthermore, such "tailed" oligonucleotides could prevent the action of poly(I)·poly(C) in extracts of interferon-treated L cells and in this regard were much more active than unmodified oligomers such as p5'A2'p5'A2'p5'A.[133]

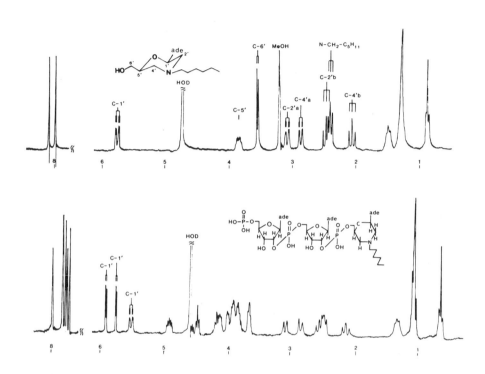

Figure 11. Proton NMR's of compounds 24 (upper) and 28 (lower).

The superior activity of compds 29 and 28 as antagonists of 2-5A action strongly suggested the synthesis of corresponding triphosphate. The triphosphate of 29 (Figure 13), compd. 30, was generated in the usual manner via pyrophosphate displacement on the corresponding imidazolide. Phosphorus NMR showed three resonances of internucleotide phosphate residues and two doublets (-5.62 ppm, J = 19 Hz, γP and -10.69 ppm, J = 19 Hz, αP) and a triplet (-20.74 ppm, J = 19 Hz, βP) (Figure 14). When this "tailed" tetramer triphosphate, 30, was evaluated for its ability to inhibit translation, it was

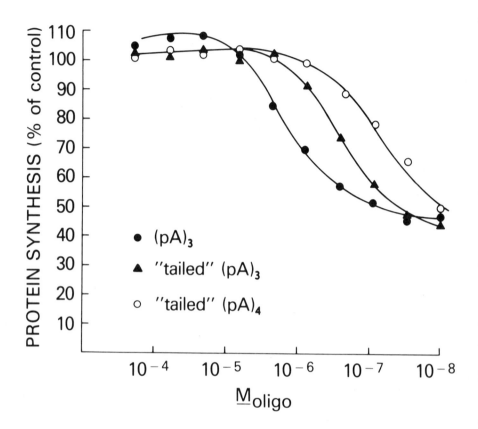

<u>Figure 12</u>. Prevention of 2-5A action by p5'A2'p5'A2'p5'A (●) or the "tailed" analogues

found that it was 10 times more active than 2-5A trimer triphosphate as an inhibitor of protein synthesis (Figure 15). In addition "tailed" tetramer triphosphate, <u>30</u>, enhanced degradation of labeled viral RNA with a concentration of $9 \cdot 10^{-11} \underline{M}$ necessary to achieve a half-maximal effect compared to 2-5A triphosphate which needs a concentration of $9 \cdot 10^{-10} \underline{M}$ to elicit a half-maximal response.

In order to explore the possibility that the enhanced activity of the "tailed" oligonucleotide, <u>i.e.</u>, <u>28</u>, <u>29</u>, and <u>30</u>, may be due to increased resistance to degradation by

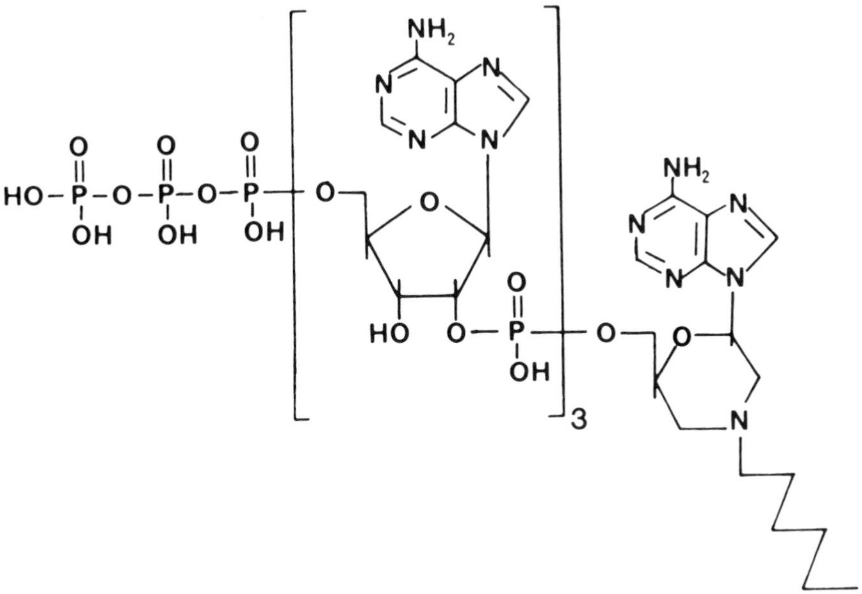

Figure 13. Structure of "tailed" 2-5A tetramer triphosphate, 30.

various enzyme activities in the cell extract, the degradation of both modified and unmodified oligomers was examined. In one test, the relative inactivation of the biological activity of 2-5A trimer triphosphate compared to the "tailed" tetramer triphosphate 30 was studied in a mouse L cell-free system under protein synthesis conditions. As Figure 16 clearly shows the modified tetramer triphosphate, 30, was completely stable for at least five hours but the unmodified oligomer ppp5'A2'p5'A2'p5'A was destroyed rapidly with a half-life of 15 minutes. Similar results have been obtained using an immunoenzymometric assay for 2-5A.[133]

Paul F. Torrence *et al.*

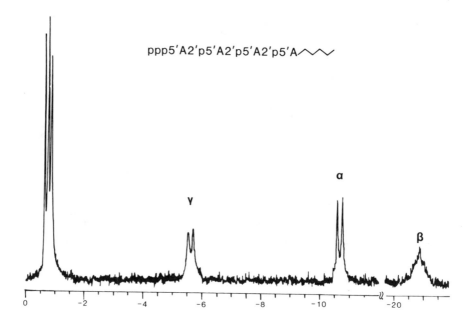

ppp5'A2'p5'A2'p5'A2'p5'A

Figure 14. ^{31}P NMR of "tailed" 2–5A tetramer triphosphate, 30.

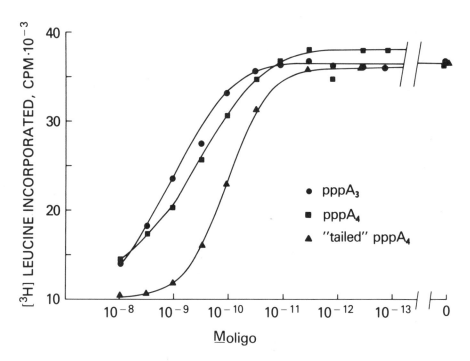

Figure 15. Inhibition of protein synthesis by 2-5A trimer triphosphate (●), 2-5A triphosphate (■), or "tailed" 2-5A tetramer triphosphate (30, ▲) in extracts of mouse L cells.

Thus, it appears that intact terminal ribose moiety is responsible for degrading 2-5A (Figure 17). This enzyme is probably similar to other phosphodiesterases in so far as modification of the 2'(3') terminal ribose hydroxyl group usually retards the rate of degradation.[134] In related studies, Baglioni et al.[116] found that that the terminally methylated analogue, ppp5'A2'p5'A2'p5'A$_m$, was considerably more stable than 2-5A in cell extracts, and Silvermann et al.[110] described 3'-terminally phosphorylated 2-5A analogues which were more stable than 2-5A itself. Furthermore, extensive

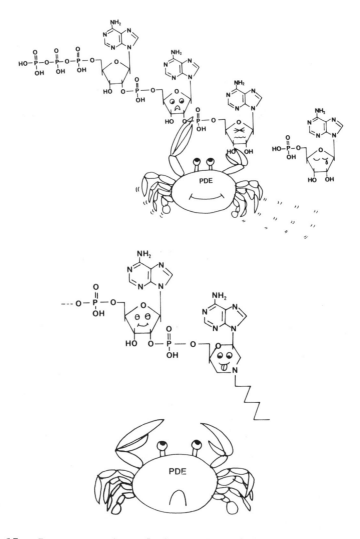

Figure 17. Representation of the action of the 2–5A phosphodiesterase (PDE) and the effect of the "tailing" modification
to the 2–5A molecule. In the upper figure, the PDE is
able to degrade the 2–5A molecule in stepwise fashion since
there exists a terminal ribose moiety and associated hydroxyl
groups for the enzyme to bind to. In the lower picture, the
PDE is unable to attack the "tailed" 2–5A since the terminal
ribose hydroxyls are missing.

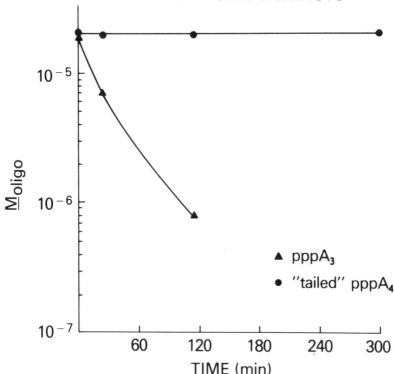

Figure 16 . Stability of 2-5A (▲) or "tailed" tetramer
triphosphate (30, ●) in extracts of mouse L cells.

modification of the terminal ribose ppp5'A2'p5'A2'p5'A2'p5'A
is possible without adversely affecting binding or activation
of the 2-5A-dependent endonuclease.   Insofar as inhibition of
protein synthesis and activation of enhanced RNA degradation
is concerned, the "tailed" tetramer triphosphate, 30, is the
most active 2-5A derivative reported to date. Modified derivatives such as 30 are readily derived from 2-5A and its
congeners; moreover, the nature of the modification may permit
further extensive alterations of the 2'-terminus of the 2-5A
molecule.

Earlier we defined two problems associated with any
envisioned chemotherapeutic applications of the 2-5A system.
The first involved the destruction of the 2-5A molecule by
various degradative enzymes, such as the 2-5A phosphodiesterase. The second problem was related to the highly charged
nature of 2-5A and its subsequent cellular impermeability.
Through synthesis and biological evaluation of various
analogues of 2-5A, it has become possible to define at least
one site in the molecule which can be altered without adversely
affecting binding to the 2-5A-activated endonuclease. One
modification executed at this site, specifically, conversion
of the 2'-terminal ribose to an N-substituted morpholine ring,
results in an 2-5A analogue which is 10 times more active as
an inhibitor of translation than the naturally occurring 2-5A
itself. This increased activity appears to be due to considerably enhanced resistance to degradation. Thus the first
problem defined above may be considered resolved. The second
and clearly more difficult challenge, that of obtaining a perme-
able form of 2-5A, remains to be met.

References

1. Isaacs, A. and Lindenmann, J. (1957) Proc. R. Soc London Ser.
   B. B147, 258-267.
2. Kazar, J., Krantwurst, P. A. and Gordon, F. B. (1971)
   Infect. Immun. 3, 819-823.
3. Beck, G., Poindrou, P., Illinger, D., Beck, J. P., Ebel,
   J. P. and Falcoff, R. (1974) FEBS Lett 48, 297-299.
4. Nebert, D. W. and Friedman, R. M. (1973) J. Virol. 11,
   193-197.
5. Cantell, K. (1979) in Interferon 1, ed., Gresser, I.,
   Academic Press N. Y., pp. 1-28.
6. Paucker, K., Cantell, K. and Henle, W. (1962) Virology 17,
   324-334.

7. Isaacs, A. Burke, D. C. (1958) Nature 182, 1073-1074.

8. Stewart II, W. E., DeClercq, E., Billiau, A., Desmyter, J. and De Somer, P. (1972) Proc. Natl. Acad. Sci. USA 69, 1851-1854.

9. Kohn, L. D., Friedman, R. M. Holmes, J. M. and Lee, G. (1976) Proc. Natl. Acad. Sci. USA 73, 3695-3699.

10. Brouty-Boye, D. and Torey, M. G. (1977) Intervirology 9 245-252.

11. Herberman, R. B., Ortaldo, J. and Bonnard, G. (1979) Nature 277, 221-223.

12. Keay, A. and Grossberg, S. E. (1980) Proc. Natl. Acad. Sci. USA 77, 4099-4103.

13. Rossi, G. B., Dolei, A., Croe, L., Benedetto, A., Matarese, G. P. and Belardelli, F. (1977) Proc. Natl. Acad. Sci 74, 2036-2040.

14. Fitzpatrick, F. A. and Stringfellow, P. A. (1980) J. Immunol. 125, 431-437.

15. Lin, S. L., Ts'o P. O. P. and Hollenberg, M. D. (1980) Biochem. Biophys. Res. Commun. 97, 168-174.

16. Chandrabose, K., Cuatrecasas, P. and Pottathil, R. (1981) Biochem. Biophys. Res. Commun. 98, 661-668.

17. Taniguichi, T., Sakai, M., Fujii-Kuriyama, Y., Muramatsu, M. Kobayashi, S. and Sudo, T. (1979) Proc. Japan Acad. 55, B, 464-469.

18. Nagata, S., Taira, H., Hall, A., Johnsrud, L., Streuli, M., Escodi, J., Boll, W., Cantell, K. and Weissman, C. (1980) Nature 284, 316-320.

19. Taniguchi, T., Fujii-Kiruyama, Y., and Muramatsu, M. (1980) Proc. Natl. Acad. Sci. USA 77, 4003-4006.

20. Derynck, R., Content, J., De Clercq, E., Volchaert, G., Tavernier, J., Devos, R. and Fiers, W. (1980) Nature 285, 542-547.

21. Derynck, R., Remaut, E., Saman, E., Staussens, P., De Clercq, E., Content, J. and Fiers, W. (1980) Nature 287, 193-197.

22. Taniguichi, T., Guarente, L., Roberts, T. M., Kimelman, D., Douham III, J. and Ptashne, M. (1980) Proc. Natl. Acad. Sci. USA 77, 5230-5233.

23. Maeda, S., McCandless, R., Gross, M., Sloma, A., Familleti, P. C., Tabor, J. M.,Evinger, M., Levy, W. P. and Pestka, S. (1980) Proc. Natl. Acad. Sci. USA 77, 7010-7013.

24. Weissman, C. (1982) in Interferon 3, ed., J. Gresser Academic Press, N. Y. 101-134.

25. Goeddel, D. V., Yelverton, E., Ullrich, A., Heyneker, H. L., Miozzari, G., Holmes, W., Seeburg, P. H., Dull, T. May, L., Stebbing, N., Crea, R., Maeda, S., McCandless, R., Sloma, A., Tabor, J. M., Gross, M., Familletti, P. C. and Pestka, S. (1980) Nature 287, 411-417.

26. Gray, P. W., Leung, D. W., Pennua, D. Yelverton, E., Najarian, R., Simonsen, C. C., Derynck, R., Sherwood, P. J., Wallace, D. W., Berger, S. L., Levanson, A. D. and Goeddel, D. V. (1982) Nature 295, 503-508.

27. Gray, P. W. and Goeddel, D. V. (1982) Nature 299, 859-863.

28. Rubinstein, M., Levy, W. P., Moschera, J. A., Lai, C.-Y., Hershberg, R. D., Bartlett, R. T. and Pestka, S. (1981) Arch, Biochem Biophys. 210, 307-318.

29. Zoon K. C., Smith, M. E., Bridgen, P. J., Zur Nedden, D. and Anfinsen, C. B. (1979) Proc. Natl. Acad. Sci. USA 76, 5601-5605.

30. Zoon, K. C., Smith, M. E., Bridgen, P. J., Anfinsen, C. B., Hunkapiller, M. W. and Hood, L. E. (1980) Science 207, 527-528.

31. Rubinstein, M., Rubinstein, S., Familletti, P. C., Mitler, R. S., Waldman, A. A. and Pestka, S. (1979) Proc. Natl. Acad. Sci. USA 76, 640-644.

32. Secher, D. S. and Burke, D. C. (1980) Nature 285, 446-450.

33. Taira, H. Broeze, R. J., Jayaram, B. M., Lengyel, P., Hunkapiller, M. W., and Hood L. E. Science (1980) 207, 528-529.

34. Knight, E. Jr., Hunkapiller, M. W., Korant, B. D., Hardy, R. W. F. and Hood L. E. (1979) Science 207, 525-526.

35. Allen, G. and Fantes, K. H. (1980) Nature 287, 408-411.

36. Levy, W. P., Rubinstein, M., Shirely, J., Del Valle, U., Lai, C.-V., Moschera, J., Brink, L., Gerber, L., Stein, S. and Pestka, S. (1981) Proc. Natl. Acad. Sci. USA 78, 6186-6190.

37. Miller, D. L., Kung, H.-F. and Pestka, S. (1982) Science 215, 689-690.

38. Gutterman, J. V., Fine, S., Quesada, J., Horning, S. J., Levine, J. F., Alexanian, R., Berhnardt, L., Kramer, M., Spiegel, H., Colburn, W., Trown, P., Merigan, T. and Dziewanowski, Z. (1982) Ann. Internal. Med. 96, 549-556.

39. Nagata, S., Mantei, N. and Weissmann, C. (1980) Nature 287 401-408.

40. Streuli, M., Nagata, S. and Weissman, C. (1980) Science 209, 1343-1347.

41. Yip, Y. K., Barrowclough, B. S., Urban, C. and Vilcek, J. (1982) Science 215, 411-412.

42a. Kaplan, N. O. and Slimmer, S. (1981) in Cellular Responses to Molecular Modulation, (Mozes, L. W., Schultz, J., Scott, W. A. and Werner, R. eds.), Academic Press, N. Y. pp. 443-453.

42b. Pestka, S., Maeda, S., Hobbs, D. S., Levy W. P., McCandless, R., Stein, S., Moeschera, J. and Staehelin, T. (1981) in Ibid, pp. 455-489.

43c. Weck, P. K., Apperson, S., Stebbing, N., Gray, P. W., Leung, D., Shepard, H. M. and Goeddel, D. V. (1981) Nucleic Acids Res. 9, 6153-6166.

44. Torrence, P. F. and De Clercq, E. (1981) Methods Enzymol. 78, 291-299.

45. Friedman, R. M. (1979) in Interferon 1. ed., Gresser, I., Academic Press N. Y., pp. 29-51.

46. Revel, M. (1979) in Interferon 1 ed., Gresser, I., Academic Press, N. Y., pp. 101-163.

47. Friedman, R. M. (1977) Bact. Rev. 41, 543-567.

48. Torrence, P. F. (1982) Mol. Aspects Med. 5, 129-171.

49. Sen. G. C. (1982) Progr. Nucleic Acid Res. Mol. Biol. 27, 105-156.

50. Lengyel, P. (1982) in Interferon 3. Gresser, J., I., ed., Academic Press, N. Y. pp. 77-99.

51. Kerr, I. M., Brown, R. E. and Ball, L. A. (1974) Nature 250, 57-59.

52. Friedman, R. M., Metz, D. H., Esteban, R. M., Tovell, D. R., Ball, L. A. and Kerr, I. M. (1972) J. Virol. 10, 1184-1198.

53. Roberts, W. K., Clemens, M. J. and Kerr, I. M. (1974) Proc. Natl. Acad. Sci. USA 73, 3136-3146.

54. Hovanessian, A. G., Brown, R. E. and Kerr, I. M. (1977) Nature 268, 537-539.

55. Kerr, I. M. and Brown, R. E. (1978) Ann. Natl. Acad. Sci. USA 75, 256-260.

56. Torrence, P. F., Johnston, M. I., Epstein, D. A., Jacobsen, H. and Friedman, R. M. (1981) FEBS Lett 130, 291-296.

57. Baglioni, C., Minks, M. A. and De Clercq, E. (1981) Nucleic Acids Res. 9, 4939-4950.

58. Dougherty, J. P., Samanta, H., Farrell, P. J. and Lengyel, P. (1980) J. Biol. Chem. 255, 3813-3816.

59. Samanta, H., Dougherty, J. P. and Lengyel, P. (1980) J. Biol. Chem. 255, 9807-9813.

60. Floyd-Smith, G., Slattery, E. and Lengyel, P. (1981) Science 212, 1030-1032.

61. Wreschner, D. H., McCauley, J. W., Skehel, J.J. and Kerr, I. M. (1981) Nature (London) 289, 414-417.

62. Williams, B. R. G., Golgher, R. R., and Kerr, I. M. (1978) FEBS Lett. 105, 47-52.

63. Hovanessian, A. G. and Wood, J. N. (1980) Virology 101, 81-90.

64. Higashi, Y. and Sokawa, Y. (1982) J. Biochem. (Tokyo) 91, 2021-2028.

65. Williams, B. R. G., Golgher, R. R., Brown, R. E., Gilbert, C. S. and Kerr, I. M. (1979) Nature (London) 282, 582-586.

66. Nilsen, T. W., Maroney, P. A. and Baglioni, C. (1982) J. Virol. 42, 1039-1045.

67. Wreschner, D. H., James, T. C., Silverman, R. H., and Kerr, I. M. (1981) Nucleic Acids Res. 9, 1571-1581.

68. Zilberstein, A., Federman, P., Shulman, L. and Revel, M. (1976) FEBS Lett. 68, 119-124.

69. Lebleu, P., Sen, G. C., Shaila, S., Carber, B. and Lengyel, P. (1976) Proc. Natl. Acad. Sci. USA 73, 3107-3111.

70. Shaila, S., Lebleu, B., Brown, G. E., Sen, G. C. and Lengyel, P. (1977) J. Gen. Virol 37, 535-546.

71. Samuel, C. E. (1979) Virology 93, 281-285.

72. Vandenbussche, P., Content, J., Lebleu, B. and Werenne, J. (1979) J. Gen. Virol 41, 161-166.

73. Lewis, J. A., Falcoff, E. and Falcoff, R. (1978) Eur. J. Biochem. 86, 497-509.

74. Kimchi, A., Zilberstein, A., Schmidt, A., Shulman, L. and Revel, M. (1979) J. Biol. Chem. 254, 9846-9853.

75. Epstein, D. A., Torrence, P. F. and Friedman, R. M. (1980) Proc. Natl. Acad. Sci. USA 77, 107-111.

76. Jagus, R., Anderson, W. F. and Safer, B. (1981) Progr. Nucleic Acid Res. Mol. Biol. 25, 127-185.

77. Safer, B., and Jagus, R. (1981) Biochimie 63, 809-817.

78. Baglioni, C. and Maroney, P. A. (1981) Biochemistry 20, 758-762.

79. Kuroda, Y., Merrick, Y. W. C. and Sharma, R. K. (1981) Science 215, 415-416.

80. Torrence, P. F., Imai, J. and Johnston, M. I. (1981) Proc. Natl. Acad. Sci USA 78, 5993-5997.

81. Knight, M., Cayley, P. J., Silverman, R. H., Wreschner, D. H., Gilbert, C. S., Brown, R. E. and Kerr, I. M. (1980) Nature 288, 189-197.

82. Torrence, P. F., Imai, J., Lesiak, K., Johnston, M. I. Jacobsen, H., Friedman, R. M., Sawai, H., and Safer, B.

(1982) in <u>Interferons</u> UCLA Symposia on Molecular and
Cellular Biology, (Merigan, T., Friedman, R. and Fox, C.
F., eds) vol XXV, Academic Press, N. Y., 123–142.

83. Jacobsen, H., Epstein, D. A., Friedman, R. M. and Torrence,
    P. F. Proc. Natl. Acad. Sci. USA, in press.

84. Janik, B., Kotick, M. P., Kreiser, T. H., Reverman, L. F.,
    Sommer, R. G. and Wilson, D. P. (1972) Biochem. Biophys. Res.
    Commun. <u>46</u>, 1153–1160.

85. Nilsen, T. W., McCandless, S. and Baglioni, C. (1982) Virolo-
    gy in press.

86. Epstein, D. A., Czarniecki, C. W., Jacobsen, H., Friedman,
    R. M. and Panet, A. (1981) Eur. J. Biochem. <u>118</u>, 1–15.

87. Robbins, C. H., Kramer, G., Saneto, R., Hardesty, B. and
    Johnson, H. M. (1981) Biochem. Biophys. Res. Commun. <u>103</u>,
    103–110.

88. Samuel, C. E. and Knutson, G., S. (1981) Virology <u>114</u>,
    302–306.

89. Gupta, S. L. (1979) J. Virology <u>29</u>, 301–311.

90. Verhaegen, M. Divizia, M., Vandenbussche, P., Kuwata, T. and
    Content, J. (1980) Proc. Natl. Acad. Sci USA <u>77</u>, 4479–4483.

91. Hovanessian, A. G., Meurs, E. and Montagnier, J. (1981) J.
    Interferon Res. 1, 179–190.

92. Shimizu, N. and Sokawa, Y. (1979) J. Biol. Chem. <u>254</u>, 12034–
    12037.

93. Johnston, M. I., Zoon, K. C., Friedman, R. M. De Clercq, E.
    and Torrence, P. F. (1980) Biochem. Biophys. Res. Commun.
    <u>97</u>, 375–383.

94. Hovanessian, A. G. and Kerr, I. M. (1978) Eur. J. Biochem.
    <u>84</u>, 149–159.

95. Etienne-Smekens, M., Vassart, G., Content, J. and Dumont, J.
    E. (1981) FEBS Lett <u>125</u>, 146–150.

96. Nilsen, T. W., Wood, D. L. and Baglioni, L. (1981) J. Biol.
    Chem. <u>256</u>, 10751–10754.

97. Stark, G. R., Dower, W. J., Schimke, R. T., Brown, R. E. and Kerr, I. M. (1979) Nature 278, 471–473.

98. Krishnan, I. and Baglioni, C. (1980) Proc. Natl. Acad. Sci. USA 77, 6506–6510.

99. Oikarinen, J. (1982) Biochem. Biophys. Res. Commun. 105, 876–881.

100. Besancon, F., Bourgeade, M. F., Justesen, J., Ferbus, D. and Thang, M. N. (1981) Biochem. Biophys. Res. Commun. 103, 16–24.

101. Kimchi, A. (1981) J. Interferon Res. 1, 559–569.

102. Kimchi, A., Shure, H. and Revel, M. (1981) Eur. J. Biochem. 114, 5–10.

103. Revel, M., Kimchi, A., Shulman, L., Wolf, D., Merlin, G., Schmidt, A., Friedman, M. Lapidot, Y. and Rapoport, S. (1981) in Cellular Responses to Molecular Mediators, eds., Mozes, L. W., Schultz, J., Scott, W. A. and Werner, R., Academic Press, New York, pp. 361–384.

104. Gresser, I., Tovey, M. G., Maury, C. and Chouroulinkov, I. (1975) Nature (London) 258, 76–78.

105. Gresser, I., Morel-Maroger, L., Riviere, Y., Guillon, J.-C., Tovey, M. G., Woodrow, D., Sloper, J. C. and Moss, J. (1980) Ann. N. Y. Acad. Sci. 350, 12–20.

106. Gresser, I., Maury, C., Tovey, M. G. Morel-Maroger, L. and Pontillon, F. (1976) Nature (London) 263, 420–422.

107. Schmidt, A., Zilberstein, A., Shulman, L., Federman, P., Berissi, H. and Revel, M. (1978) FEBS Lett 95, 257–264.

108. Williams, B. R. G., Kerr, I. M., Gilbert, C. S., White, C. N., and Ball, L. A. (1978) Eur. J. Biochem. 92, 455–462.

109. Schmidt, A., Chernajovsky, L., Shulman, P., Federman, P., Berissi, H. and Revel, M. (1979) Proc. Natl. Acad. Sci. USA 76, 4788–4792.

110. Silverman, R. H., Wreschner, D. H., Gilbert, C. S. and Kerr, I. M. (1981) Eur. J. Biochem. 115, 79–85.

111. Lesiak, K., Imai, J. and Torrence, P. F., unpublished observations.

112. Sawai, H., Shibata, T. and Ohno, M. (1981) Tetrahedron 37, 481–485.

113. Cramer, F., Schaller, H. and Staab, H. A. (1961) Chem. Ber. 94, 1612– 1621.

114. Hoard, D. E. and Ott, D. G. (1965) J. Amer. Chem. Soc. 87, 1785–1788.

115. Richardson, C. (1972) in Procedures in Nucleic Acid Research (G. L., Cantoni and D. R. Davies, eds.) Harper and Row, New York, pp. 815–828.

116. Baglioni, C. D'Allessandro, S. B., Nilsen, T. W., den Hartog, J. A. J., Crea, R. and Van Boom, J. H. J. Biol. Chem. 256, 3253–3257 (1981).

117. Schleich, T., Cross, B. P. and Smith, I. C. P. (1976) Nucleic Acids Res. 3, 355–370.

118. Kondo, N. S., Holmes, H. M., Stempel, L. M. and Ts'O, P. O. P. (1970) Biochemistry 9, 3479–3498.

119. Dhingra, M. M. and Sarma, R. H. (1978) Nature 272, 798–801.

120. Doetsch, P., Wu, J. M., Sawada, Y. and Suhadolnik, R. J. (1981) Nature 291, 255–258.

121. Doetsch, P. W., Suhadolnik, R. J., Sawada, Y., Mosca, J. D., Flick, M. B., Reichenbach, N. L., Dang, A. Q., Wu, J. M., Charubala, R., Pfleiderer, W. and Henderson, E. E. (1981) Proc. Natl. Acad. Sci. USA 78, 6699–6703.

122. Charubala, R. and Pfleiderer, W. (1980) Tetrahedron Lett 21, 4077–4080.

123. Sawai, H., Imai, J., Lesiak, K., Johnston, M. I. and Torrence, P. F., J. Biol. Chem. in press.

124. Suhadolnik, R. J., Doetsch, P., Wu, J. M., Sawada, Y., Mosca, J. D. and Reichenbach, N. L. (1981) Methods Enzymol. 79, 257–265.

125. Sawai, H. and Ohno, M. (1981) Bull. Chem. Soc. Jpn. 54, 2759–2762.

126. Sawai, H. and Ohno, M. (1981) Chem. Pharm. Bull 29, 2237–2245.

127. Lesiak, K. and Torrence, P. F., unpublished observations.

128. Lesiak, K. and Torrence, P. F., submitted for publication.

129. Drocourt, J.-L., Dieffenbach, C. W., Ts'o, P. O. P., Justensen, T. and Thang, M. N. (1982) Nucleic Acids Res. 10, 2163–2174.

130. Khym, J. C. (1963) Biochemistry 2, 344–350.

131. Brown, D. M. and Read, A. P. (1965) J. Chem. Soc. 5072–5074.

132. Erlanger, B. F., Methods Enzymol. 70, 85–104 (1980).

133. Imai, J., Johnston, M. I. and Torrence, P. F., J. Biol. Chem., in press.

134. Sierakowska, H. and Shugar, D. (1977) Progr. Nucleic Acid Res. Mol. Biol. 20, 60–130.

RECEPTORS FOR ADENOSINE AND ADENINE NUCLEOTIDES

Robert F. Bruns

Department of Pharmacology
Warner-Lambert/Parke Davis Pharmaceutical Research
Ann Arbor, Michigan

Responses to adenosine (ado) include coronary vaso-
dilation, reduction in heart rate and force, inhibition
of platelet aggregation, inhibition of lipolysis, renal
vasoconstriction, and sedation. Conversely, the xanthines caffeine and theophylline block ado receptors and
have effects such as increases in heart rate and force,
lipolysis, diuresis, and locomotor stimulation. Although
these effects are opposite to effects of ado, they have
not been conclusively shown to be due to blockade of
endogenous ado. Responses to ado are mediated by ado
receptors, which have been divided into A_1 and A_2
subtypes, causing decreases and increases in adenylate
cyclase activity, respectively. When 128 nucleosides were
tested in human fibroblasts (A_2 response), 14 were full
agonists and 11 were antagonists. Among the antagonists
were three cycloados which are locked in anti conformation. ^{3}H-N^6-cyclohexyladenosine (^{3}H-CHA) binds to A_1
receptors in brain membranes (K_D ~1 nM, B_{max} ~50
pmol/g wet weight). ^{3}H-CHA binding is regulated by ions
and guanine nucleotides and inhibited competitively by
nucleosides which are active at A_1 receptors and by
xanthine ado antagonists such as theophylline. The
antagonist ^{3}H-1,3-diethyl-8-phenylxanthine (^{3}H-DPX)
also binds to A_1 receptors but its affinity is highly
species-dependent. In striatum, ^{3}H-N-ethyladenosine
5'-carboxamide (^{3}H-NECA) binds to the same A_1 receptor
as ^{3}H-CHA and ^{3}H-DPX, but in addition binds to a second
ado receptor which may be A_2. Potencies of different
xanthines in increasing locomotor activity in mice
correlate with affinities in ^{3}H-CHA binding.

Although AMP, ADP, and ATP cause responses via ado
receptors, these responses require hydrolysis to ado
because they are blocked by 5'-nucleotidase inhibitors
and ado deaminase. Thus adenine nucleotides do not
interact with the ado receptor. Conversely, there exist
receptors for ADP and ATP which do not recognize ado and
are not blocked by xanthines.

I.   INTRODUCTION

In the half-century since adenosine's effects on the
cardiovascular system were reported (1), many other effects
of the nucleoside have been discovered. These include
increases in coronary blood flow, negative inotropic and
chronotropic responses, inhibition of lipolysis, inhibition
of platelet aggregation, renal vasoconstriction, and behavioral sedation (2,3). Biochemically, adenosine increases
adenylate cyclase activity and cyclic AMP levels in brain
slices (4) and many cultured cell lines (5,6), but decreases
adenylate cyclase in fat cells (7) and primary glial cultures
(8). Other second messengers such as calcium (9) are also
probably regulated by adenosine.

Despite the length of time that many of these effects have
been known, it was not until the early 1970s that the concept
of receptors for adenosine began to appear in the literature.
We now know that responses to adenosine are mediated by specific adenosine receptors, which are similar in their properties
to the well-known receptors for catecholamines, acetylcholine,
peptide hormones, etc. There are several pieces of evidence
which support the adenosine receptor concept. The response
to adenosine is very specific, requiring both the adenine and
ribose parts of the molecule. There is a high degree of
stereospecificity: inverting any one of the four asymmetric
carbons of adenosine results in a completely inactive compound. Many other simple changes in the adenosine structure,
such as 2'-deoxy or 7-deaza modification, also drastically
reduce activity. The biochemical consequences of adenosine
receptor activation (increases or decreases in cyclic AMP)
are reminiscent of other receptors. For instance, β-adren-
ergic receptors cause increases in cyclic AMP, while α_2-
adrenergic receptors inhibit cyclic AMP accumulations. Like

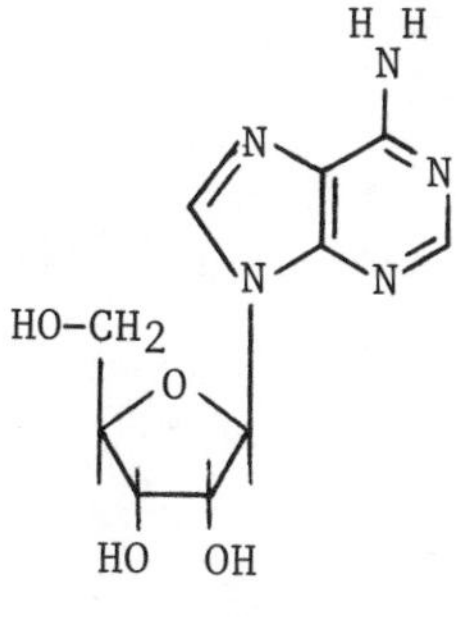

adenosine

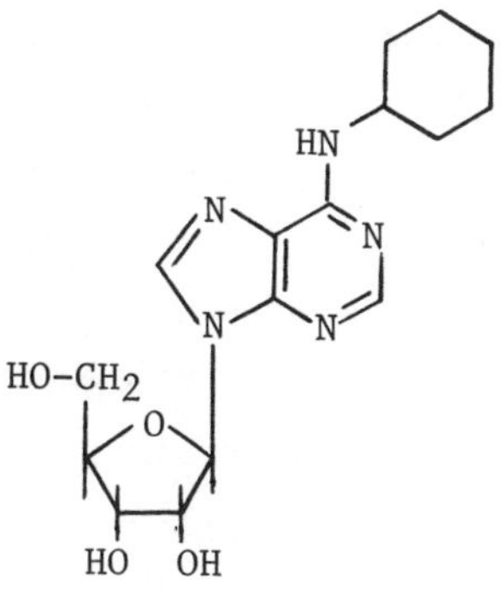

CHA

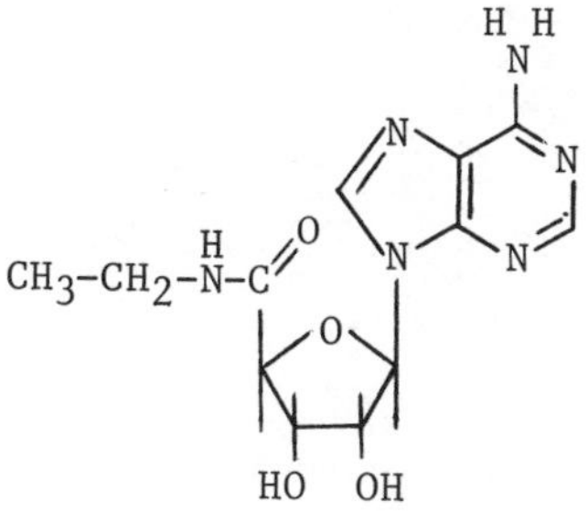

NECA

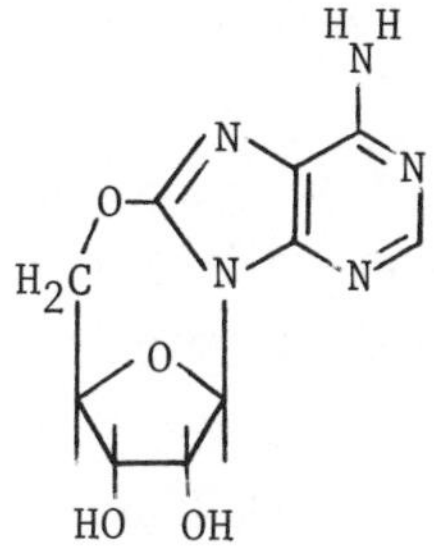

8,5'-O-cycloadenosine

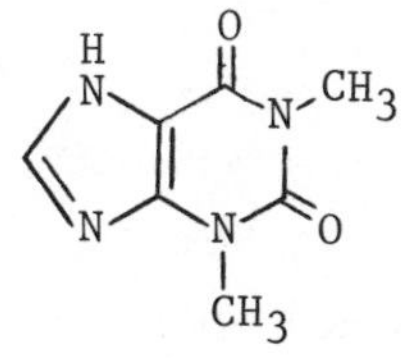

theophylline

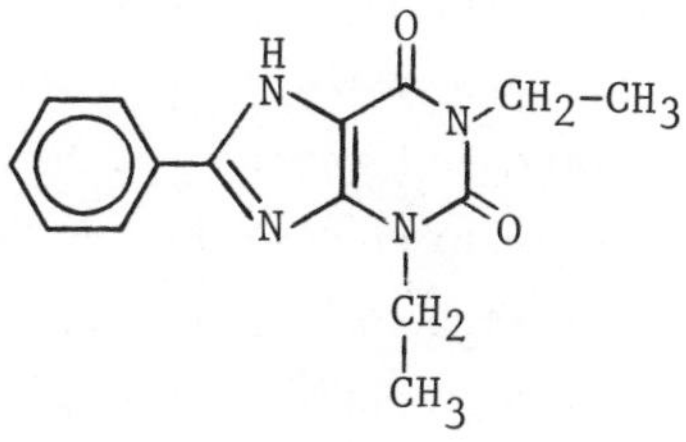

DPX

FIGURE 1. Structures of compounds which interact with adenosine receptors. Abbreviations: CHA, N^6-cyclohexyladenosine; NECA, N-ethyladenosine-5'-carboxamide; DPX, 1,3-diethyl-8-phenylxanthine.

other receptors, adenosine receptors have antagonists. The
methylxanthine theophylline (Fig. 1) has long been known to
competitively block the response to adenosine, much as
propranolol blocks responses to norepinephrine. Like other
receptors, the adenosine receptor is extracellular, since
adenosine uptake blockers potentiate rather than block the
response to adenosine, and since adenosine linked to polymers
retains activity (10).

Receptors for ADP and ATP are also known to exist, and
will be discussed later in this paper.

II. ADENOSINE RECEPTOR STRUCTURE-ACTIVITY RELATIONS: NUCLEOSIDES

In order to better characterize the adenosine receptor,
128 nucleosides were tested for adenosine agonist and
antagonist activity in VA13 fibroblasts (11), an SV40 virus
transformed cell line derived from WI-38 human fetal lung
fibroblasts. VA13 cells respond to adenosine (EC_{50} 15 µM)
with up to 50-fold increases in cyclic AMP. Isoproterenol and
prostaglandin E_1 cause even larger cyclic AMP increases, which
can be used as controls to test for specificity of blockers.
Of the 128 analogs of adenosine tested, 55 were completely
inactive and another 20 had only very weak activity, confirming
the very high specificity of the receptor. Fourteen compounds
were full agonists (i.e., caused a maximal response equal to
adenosine) and only two were more potent than adenosine.
N-ethyladenosine-5'-carboxamide (NECA) (from Dr. Karl Dietmann)
was five times as potent as adenosine, and 2-azaadenosine (from
Dr. John Montgomery) was twice as potent as adenosine.

TABLE I. Low-Efficacy[a] Partial Agonists

Compound	Apparent K_i µM	Efficacy, %	Source[b]
5'-0-Methyladenosine	6.2	6.2	A
5'-Deoxy-5'-iodoadenosine	8.0	4.3	B
Adenosine 5'-acetate ester	21	c	A
Adenosine 5'-nitrate ester	9.0	c	C
5'-Deoxy-5'-[(ethoxycar-bonyl)methyl]adenosine	255		A
Adenosine-5'-carboxylic acid, ethyl ester	5.9	5.2	D
9-(3-deoxy-β-D-erythro-pent-3-enofuranosyl)-adenine	150	20	E

[a]Efficacy was defined as the maximal response to the compound
as a percentage of the maximal response to adenosine. Low-
efficacy partial agonists were defined as compounds with
efficacies less than 30% which were specific competitive
inhibitors of the response to adenosine. Values are means of
"best fit" parameter values from separate experiments.
Adapted from (11).
[b]Sources: A, Dr. H. Follmann; B, Aldrich Chemical; C, Dr. R.
Duschinsky; D, Dr. J. W. Daly; E, Dr. J. G. Moffatt.
[c]Efficacy was reduced to 1-2% by adenosine deaminase. The
compounds may be pure antagonists.

Seven compounds were partial agonists: at maximal concentrations they caused small responses alone (5-20% of the
maximal response to adenosine) and blocked the response to
adenosine (Table I). The blockade was specific for adenosine
receptors since responses to isoproterenol and prostaglandin
E_1 were not affected. Interestingly, all of the partial
agonists were 5'-modified analogs. 3'-Deoxyadenosine and
5'-deoxyadenosine were also partial agonists, as previously
shown by Huang, et al. (12), but they were not specific
inasmuch as they blocked the responses to isoproterenol and
prostaglandin E_1 (a "P site" effect, see below). The
compounds in Table I are thus the first specific partial
agonists at the adenosine receptor.

TABLE II.   Competitive Inhibitors[a]

Compounds	Apparent K_i μM	Source[b]
5'-Deoxy-5'-(methylthio)adenosine	8.2	F
5'-Deoxy-5'-[(methoxycarbonyl)-methyl]adenosine	5.0	A
9-β-D-erythrofuranosyladenine	9.5	G
9-α-L-threofuranosyladenine	5.6	G
8,5'-0-Cycloadenosine	35	H
5'-Deoxy-8,5'-cycloadenosine	120	I
5'-Thio-8,5'-S-cycloadenosine	9.2	H
2-Chloro-5'-deoxy-5'-(methyl-thio)adenosine	15	J
2-Fluoro-5'-deoxy-5'-(ethyl-thio)adenosine	80	J
2-Chloro-5'-deoxy-5'-(ethyl-thio)adenosine	66	J
2-Amino-5'-deoxy-5'-(ethyl-thio)adenosine	115	J

[a]Competitive inhibitors were defined as compounds which did not increase cyclic AMP, but were specific, competitive inhibitors of the response to adenosine. K_i values are means of "best fit" parameters from separate experiments. Adapted from (11).
[b]Sources: F, Sigma Chemical; G, Dr. D. Murray; H, Dr. M. Ikehara; I, Dr. T. Ueda; J, Dr. J. Montgomery.

Eleven compounds caused no response by themselves, but blocked the response to adenosine (Table II). These are the first adenosine receptor antagonists with nucleoside structure reported in the literature. Like the partial agonists, all of the antagonists are modified at the 5'-position, suggesting that the part of the receptor which interacts with the 5'-position shifts in configuration when the receptor converts from antagonist to agonist conformation. The antagonists were competitive and did not block the responses to isoproterenol and prostaglandin E_1.

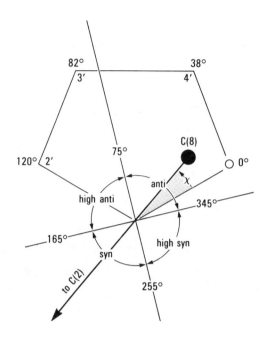

FIGURE 2. Glycosidic bond angle (χ). The point of
view is looking directly down the N(9)-C(1') bond. Angles
are calculated according to the assumption that the ribose
ring is a planar regular pentagon (eclipsed conformation)
and will vary depending on the exact conformation and bond
lengths. From Bruns (11).

The purine base of adenosine can change its position
relative to the ribose ring by rotating around the glycosidic
bond between N(9) and C(1'), as illustrated in Fig. 2.
Although adenosine adopts a torsion angle in the "anti"
range in the solid state (13), the "high anti" and "syn"
ranges are easily accessible in solution. By testing
several conformationally restricted analogs of adenosine,
it was possible to demonstrate that adenosine binds to its
receptor in "anti" conformation. The 8,5'-cycloadenosines
in Table II are covalently locked in the "anti" conformation,

yet they retain affinity as adenosine antagonists. In contrast, 8,2'- and 8,3'-cycloadenosines have no affinity at the adenosine receptor. By using other analogs with strong conformational preferences, it was shown that agonists also bind in "anti" conformation (11).

In contrast to the competitive antagonists, 2',5'-dideoxyadenosine (made by Dr. Leroy Townsend) was a non-competitive inhibitor which also inhibited responses to isoproterenol and prostaglandin E_1. The site which mediates this inhibition is not an extracellular adenosine receptor, but rather an intracellular allosteric site, probably on the catalytic subunit of adenylate cyclase (14). Londos, et al. (15) called this site the "P site", since it required that the purine portion of adenosine be unaltered. "P sites" should not be confused with Burnstock's "P_1" and "P_2" receptors (16), which are extracellular receptors for adenosine and adenine nucleotides, respectively.

III.   ADENOSINE RECEPTOR STRUCTURE-ACTIVITY RELATIONS:
       XANTHINES

As an adenosine antagonist, theophylline has several disadvantages: it has low potency (K_i 5 μM) and has a number of actions unrelated to adenosine such as phosphodiesterase inhibition and effects on calcium transport. Since very little structure-activity work had been done on the xanthines, 110 xanthines and related compounds were tested as adenosine antagonists in human fibroblasts (17). Two variations on theophylline gave substantial improvements in affinity. Replacing the 1,3-dimethyl groups of theophylline with 1,3-diethyl or 1,3-dipropyl increased affinity five- to eight-fold. Replacing theophylline's 8-hydrogen with phenyl increased affinity 25-fold. The first adenosine antagonist with nanomolar potency, 8-phenyltheophylline was reported by Smellie,

et al. (18) to be the most specific adenosine antagonist of
a series of xanthines. Unfortunately, 8-phenyltheophylline's
practical utility is limited by its low solubility. Adenosine antagonists with higher potency and solubility are
urgently needed in adenosine research.

IV. ADENOSINE RECEPTOR SUBTYPES

Adenosine receptors were divided by van Calker, et al.
(8) into two subtypes according to their effects on
adenylate cyclase: receptors which mediated decreases
in adenylate cyclase activity were called A_1, and receptors
which caused increases in adenylate cyclase were called A_2.
In an alternate nomenclature, Londos, et al. (19) called
these receptors R_i and R_a, respectively. The two receptors
differ in their affinities for various adenosine agonists:
at A_1 receptors, adenosine is active at low nanomolar concentrations and N^6-(L-phenylisopropyl)adenosine (L-PIA,
formal name N^6-[(R)-1-methyl-2-phenylethyl]adenosine) is more
potent than adenosine, which is more potent than NECA, while
at A_2 receptors, adenosine is active at micromolar or high
nanomolar concentrations, and NECA is more potent than adenosine, which is more potent that L-PIA (19). By the above
criteria, the receptor in human fibroblasts is an A_2 receptor.

V. DIRECT BINDING EXPERIMENTS AT ADENOSINE RECEPTORS

Since receptor-binding assays have been useful in
studying many hormones and neurotransmitters, we wished
to develop such assays for adenosine receptors. ^{3}H-
Adenosine had previously proven unsatisfactory for this
purpose. As a first ligand, we chose ^{3}H-N^6-cyclohexyl-
adenosine (^{3}H-CHA), which had been shown by Trost and Stock

(20) to be active at 1 nM in inhibiting lipolysis in fat
cells, an A_1 receptor-mediated response. ^{3}H-CHA also had
the advantage of being stable to adenosine deaminase, which

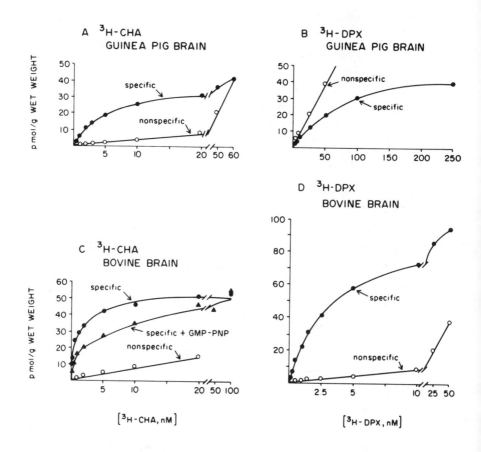

FIGURE 3.   Binding of [3H]CHA (A and C) and [3H]DPX
(B and D) in guinea pig brain (A and B) and bovine brain (C
and D).   n = 6 for total binding and 3 for nonspecific binding
and for total binding with 100 µM of GMP-PNP.   For [3H]DPX
binding in guinea pig brain (B), n = 12 for total and 9 for
nonspecific binding; in this experiment, the tissue was
preincubated for one hour at 25°C with 100 nM unlabeled CHA
to eliminate presumptive A_1 receptor binding.   From Bruns,
et al. (21).

allowed us to remove endogenous adenosine by adding adenosine
deaminase to the incubation. ^{3}H-CHA bound with high affinity
to brain membranes from guinea pig and cow (21). Specific
binding, defined as the portion of binding which could be
prevented by addition of 10 µM L-PIA, was more than 90% of
total binding. Specific binding was saturable, with maximal
binding in bovine brain of 54 pmol per g wet weight and
half-maximal binding at 0.7 nM (Fig. 3). Different adenosine
agonists and antagonists inhibited ^{3}H-CHA binding with
affinities which were consonant with their known activities
at A_1 receptors (Fig. 4). The potent A_1 agonist L-PIA had
the highest affinity, followed by 2-chloroadenosine and
N-cyclopropyladenosine-5'-carboxamide. D-PIA was about 40-
fold less active than L-PIA, which is consistent with the
known stereospecificity of the two PIA diastereomers at A_1
receptors (22). Affinities of xanthines versus ^{3}H-CHA
binding at A_1 receptors were parallel to their affinities at
A_2 receptors in fibroblasts, with 8-phenyltheophylline >
1,3-dipropylxanthine > theophylline > caffeine. This suggests
that the xanthines do not markedly differentiate between A_1
and A_2 receptors. ^{3}H-CHA binding was inhibited by the GTP
analog guanylylimidodiphosphate (GMP-PNP), which has been
shown to reduce affinities of agonists (but not antagonists)
in binding to receptors which are linked positively or
negatively to adenylate cyclase (23). Other agonist ligands
which have been used to label adenosine A_1 receptors are
[3]H-L-PIA (24) and ^{3}H-2-chloroadenosine (25,26).

We also wished to find an antagonist radioligand for
adenosine receptors. We chose ^{3}H-1,3-diethyl-8-phenyl-
xanthine (^{3}H-DPX) because it had about four-fold higher
affinity than 8-phenyltheophylline in blocking the response
to adenosine at A_2 receptors in human fibroblasts. Our
hope was that ^{3}H-DPX would bind to both A_1 and A_2 receptors.

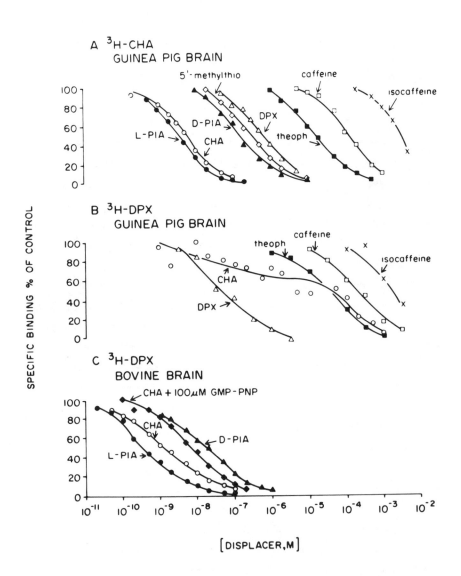

FIGURE 4. Displacement curves for nucleosides and methylxanthines. 5'-Methylthio is 5'-deoxy-5'-(methylthio)adenosine.
n = 12 for control binding, 6 for nonspecific binding, and 3
for binding in the presence of displacers. Adapted from (21).

We found that specific binding of ^{3}H-DPX was highly species-
dependent. In guinea-pig brain membranes, ^{3}H-DPX bound
saturably with low affinity (Fig. 3). About half of the
binding was displaced by nanomolar concentrations of unlabeled
CHA (Fig. 4) and probably represented A_1 receptor binding.
Although the remainder of specific binding had micromolar
affinity for nucleosides, it was probably not related to
A_2 receptor sites since the structure-activity relationship
for nucleosides as displacers of specific binding was not
consistent with an adenosine receptor. 8-Bromoadenosine,
which is inactive at adenosine receptors, was active in
displacing binding, while the potent adenosine agonist
N-cyclopropyladenosine-5'-carboxamide was very weak in
displacing binding.

In bovine brain, ^{3}H-DPX bound with high affinity
(K_D 5 nM) (Fig. 3). All of the binding was inhibited very
potently by unlabeled CHA and L-PIA (Fig. 4), indicating that
the binding was solely to A_1 receptors.

Recently, Yeung and Green (27) reported that ^{3}H-NECA
bound to two sites in rat striatal membranes. One site,
with a low nanomolar affinity for unlabeled L-PIA, undoubtedly
represents the A_1 receptor, while the other site, with an
affinity for L-PIA in the high nanomolar range, may represent
an A_2 receptor. We have reproduced these results using
unlabeled CHA to distinguish the two ^{3}H-NECA binding sites
(Fig. 5). In our hands, the dose-inhibition curve for CHA
in rat striatum is shallow, indicating two sites, but there
is no plateau to separate the two sites clearly. In bovine
striatum, there are two clearly separated sites, one having
an affinity for CHA of about 1 nM and the other about 1 μM.
In whole brain, the "A_2" site is detectable but the A_1 site
predominates, while in most peripheral tissues in rat no
binding of ^{3}H-NECA to bona fide adenosine receptors is
detectable. The exception is the testis, where the A_1

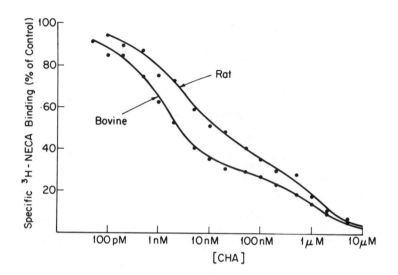

FIGURE 5.   Inhibition of ^{3}H-NECA binding to rat and
bovine striatum by unlabeled CHA. ^{3}H-NECA (4nM) was incubated at 25°C in 1 ml 50 mM tris pH 7.7 with 10 mM $MgCl_2$ for
60 min in the presence and absence of different concentrations
of CHA.   Nonspecific binding was defined as binding in the
presence of 100 nM CHA plus 1 mM theophylline.

but not the "A_2" ^{3}H-NECA binding site can be demonstrated.
A_1 receptors in testis have been described using ^{3}H-CHA
(28) and ^{3}H-2-chloroadenosine (25).   Autoradiographic
evidence indicates that the receptors are on spermatocytes
(Murphy, Goodman, and Snyder, personal communication).

VI. BEHAVIORAL ROLE OF ADENOSINE

Adenosine and its derivatives are powerful sedatives (29,
30), while xanthines such as caffeine and theophylline are
mild stimulants. This led us to hypothesize (31) that the
xanthines are stimulants by virtue of blocking the sedative
effects of endogenous adenosine, as originally suggested by
Sattin and Rall (4). In order to test this possibility, we
chose ten xanthines with widely differing affinities in ^{3}H-CHA
binding (32). The compounds also differed widely in their
effects on locomotor activity in mice (Fig. 6).

7-(β-Chloroethyl)theophylline, theophylline, and caffeine
were potent locomotor stimulants, 1,7-dimethylxanthine and 7-
(β-hydroxyethyl)theophylline were weaker stimulants, and four
other xanthines had little effect on locomotor activity.
3-Isobutyl-1-methylxanthine was a locomotor depressant. When
the threshold for locomotor stimulant activity was compared
with affinity in ^{3}H-CHA binding (Table 3), there was excellent
agreement for all compounds except 3-isobutyl-1-methylxanthine.
Since 3-isobutyl-1-methylxanthine is the only compound of the
ten which is a potent phosphodiesterase inhibitor (33), it is
possible that locomotor stimulant activity of 3-isobutyl-1-
methylxanthine due to adenosine antagonism may be masked by
locomotor depressant activity due to phosphodiesterase inhibition. On the whole, then, the data support the proposition
that the stimulant actions of methylxanthines are due to
adenosine blockade.

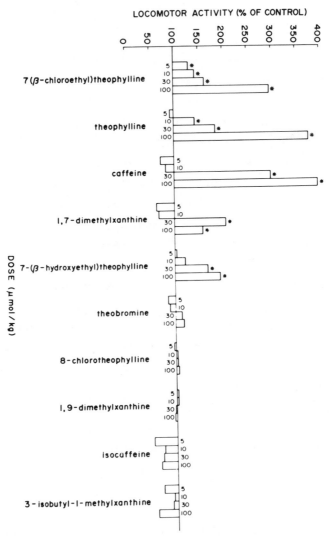

FIGURE 6.   Effect of alkylxanthines on locomotor activity
of mice.   Locomotor activity values for groups of 10-20 mice at
each dose are for the second 30 min after intraperitoneal
injections of the indicated doses except for 7-(β-chloroethyl)theophylline for which the activity values represent the
first 30-min period.   Values represent the locomotor activity
as percentage of the activity of saline-injected control mice.
Adapted from Snyder, et al. (32).
* Significantly different from saline, P < 0.005 by Student's
   t test.

TABLE III. Xanthines: Behavioral stimulant potencies and
            effects on adenosine receptor ^{3}H-CHA binding

	Xanthine	Receptor Binding IC_{50} μM	Locomotor Stim. Threshold $\mu mol/kg$
1.	7-(β-Chloroethyl)-theophylline	10	5
2.	Theophylline	23	10
3.	1,7-Dimethylxanthine	30	30
4.	3-Isobutyl-1-methyl-xanthine (IBMX)	50	>100
5.	Caffeine	50	30
6.	7-(β-Hydroxyethyl)theo-phylline	100	30
7.	Theobromine	150	>250
8.	8-Chlorotheophylline	500	>250
9.	1,9-Dimethylxanthine	>1000	>250
10.	Isocaffeine	>1000	>250

Binding of [^{3}H]CHA (1.0 nM) was assayed in triplicate with six
concentrations of xanthines. Data are means of three determin-
ations of IC_{50} values (concentration to inhibit specific
binding by 50%) which varied less than 20%. Locomotor stimulation threshold represents the minimal dose to augment monitored locomotor activity significantly (tested by statistical
analyses). For each methylxanthine, five or six doses from
2.5 to 250 $\mu mol/kg$ were evaluated with 10-20 mice at each
dose.

Vapaatalo, et al. (30) reported that L-PIA had potent

behavioral depressant activity. We found that L-PIA could

inhibit locomotor activity in mice almost completely at quite

low doses (Fig. 7). Strong depressant activity was seen at

100 nmol/kg, which means that L-PIA ranks among the most

potent psychoactive drugs, comparable to LSD and the neuro-

leptic spiperone. At the lowest active dose, L-PIA caused a

behavioral stimulation. Other depressants such as benzodia-

zepines, barbiturates, and alcohol are known to elicit low-

dose stimulation, which is thought to be a disinhibition

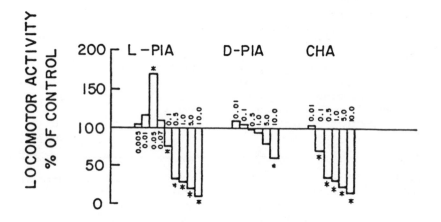

FIGURE 7.   Effects of CHA and L- or D-PIA on locomotor activity of mice.   Locomotor activity for groups of nine mice at each intraperitoneal dose for the 20- to 30-min period after drug administration are expressed as percentage of activity of saline-injected control mice.   *Significantly different from saline, P < 0.005 by Student's t test.

phenomenon. CHA was about equal to L-PIA in locomotor

depressant potency, while D-PIA was 50-fold weaker. The

very high potency of L-PIA and CHA and the lesser potency of

D-PIA imply that the locomotor depression that these compounds

exhibit is mediated by A_1 receptors. The sedation caused by

L-PIA is not due to L-PIA's hypotensive actions, since seda-

tion occurs at low doses of L-PIA which have no effect on

blood pressure (Katims, Annau, and Snyder, personal communication). The sedative actions of L-PIA are not blocked by the
putative peripheral adenosine antagonist 8-(p-sulfophenyl)-
theophylline (32).

VII. COMPARISON BETWEEN ADENOSINE RECEPTORS AND
ADENINE NUCLEOTIDE RECEPTORS

The subject thus far has been adenosine. However, many
tissues also respond to adenine nucleotides, particularly ADP
and ATP. Are these responses mediated by adenosine receptors,
or do there also exist receptors for adenine nucleotides? In
short, the answer is that both of these possibilities are true,
depending on the tissue being investigated.

The VA13 human fibroblast is an example of a system where
adenine nucleotides are active only by virtue of being converted
to adenosine (5). In VA13 cells, the increase in cyclic AMP
in response to AMP is essentially the same as the increase in
response to adenosine (Fig. 8). ADP and ATP cause smaller,
slower responses. AMP is converted to adenosine extracellularly
by the enzyme 5'-nucleotidase, which can be inhibited by purine
and pyrimidine nucleotides, the di- and triphosphates being non-
substrate inhibitors and the monophosphates being competitive
substrates. When 19 different nucleotides were tested for
their effects on the response to AMP and on the breakdown of
AMP to adenosine, a close linear relationship was seen between
the two effects (Fig. 9). The most potent compound for inhibition of both the enzyme and the response to AMP was α,β-
methylene-ADP (APCP), a non-hydrolyzable analog of ADP. These

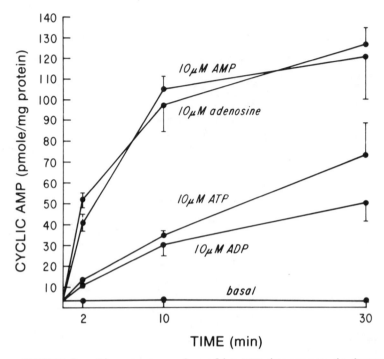

FIGURE 8. Time course of cyclic AMP increases in human fibroblasts in response to adenosine AMP, ADP, and ATP. Values are means ± S.E., n = 3. From Bruns (5).

results indicate that the response to AMP is dependent on prior conversion of AMP to adenosine. This was confirmed in experiments where addition of adenosine deaminase to the incubation eliminated the response to AMP. Since AMP is not affected by adenosine deaminase, this implies that AMP per se is incapable of activating the adenosine receptor. Responses to ADP and ATP were also blocked by adenosine deaminase.

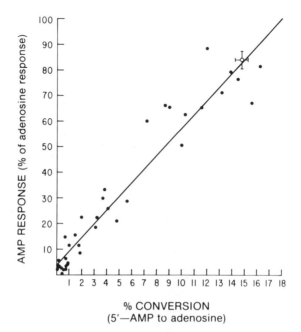

% CONVERSION
(5'—AMP to adenosine)

FIGURE 9. Effect of 5'-nucleotidase inhibitors on the
response to 10 μM AMP. Each point illustrates the mean
response and mean conversion in the presence of a single
concentration of an inhibitor. AMP alone is an open circle ±
S.E., n = 9. The line is the unweighted linear regression for
all the points. Sample sizes were usually 3 for cyclic AMP and
2 for 5'-nucleotidase. Unless otherwise stated, inhibitors
were tested at 100 μM and 1 mM. Compounds tested (listed in
decreasing order of activity vs 5'-nucleotidase) were: APCP (1
μM, 3 μM, 10 μM, 30 μM, 100 μM, 1 mM), APCPP, dTDP,
UDP, GDP, dGDP, dTTP, CDP, dUDP, GMP, dTMP, UMP, ACP, dGMP,
dUMP, 2',3'-dideoxy-TTP, β-glycerophosphate (1 mM), 2'-GMP
(1 mM), and α-D-glucose-1-phosphate (1 mM). From Bruns (5).

Finally, a number of systems with true nucleotide receptors have been described (Table IV). Mast cells degranulate
in response to ATP; ADP, AMP, and adenosine do not cause
degranulation (34). The mast cell ATP receptor is thus 100%
specific for ATP, and does not respond to other adenine-ribose
compounds, just as the adenosine receptor is 100% specific for
adenosine in the adenosine-AMP-ADP-ATP series.

TABLE IV. Purinoreceptors

| Species | Tissue | Response | Relative Potency | | | | Receptor Type(s) |
			ado	AMP	ADP	ATP	
human	VA13 fibroblast	↑ cAMP	100	0	0	0	ado
rat	mast cell	degranulate	0	0	0	100	ATP
human	platelet	aggregate	inh	inh	100	inh	ado, ADP
guinea pig	taenia coli	relax	3	3	100	100	ado, ADP?
guinea pig	bladder	contract	inh	inh	10	100	ado, ATP?

Abbreviations: ado = adenosine, inh = inhibits.

The blood platelet aggregates in response to ADP, while adenosine, AMP, and ATP inhibit aggregation (35). The inhibition by adenosine is via an A_2 adenosine receptor, which stimulates adenylate cyclase and can be blocked by theophylline (36). The platelet thus contains two different purinoreceptors. The inhibition of ADP-induced aggregation by ATP appears to be due to competitive inhibition at the ADP receptor (37), but the situation with regard to analogs of AMP and ATP is more complex (38). Macfarlane and Mills (39) used 2-methylthio-β-^{32}P-ADP to label ADP receptors in intact platelets. The platelet ADP receptor is to date the only nucleotide receptor which can be studied in direct binding experiments.

The existence of more than one purinoreceptor in the same tissue appears to be almost the rule rather than the exception. Tissues other than the platelet which have more than one purinoreceptor include the guinea pig taenia coli and the guinea pig bladder. In addition, mast cells are thought to have an A_1 adenosine receptor which potentiates the effects of degranulating agents but does not cause degranulation by itself (40).

In the guinea pig taenia coli, adenosine and all of the adenine 5'-nucleotides cause relaxation. However, the relaxations caused by ADP and ATP are qualitatively different than the relaxations caused by adenosine and AMP (41). The relaxations due to ADP and ATP occur at lower concentrations and are more rapid than the relaxations to adenosine and AMP. In addition, the non-hydrolyzable ADP analog APCP is as potent as ADP and ATP (41,42), while ACPPP, the analog of ATP in which the 5'-oxygen has been replaced with CH_2, is more potent than ADP or ATP (42). Theophylline blocks the responses to adenosine and AMP, but not to ADP or ATP (43). Although 8-bromoadenosine and 9-β-D-arabinofuranosyl-adenine are completely inactive, their triphosphates are

almost as active as ATP (44). The taenia coli therefore
clearly contains an adenosine receptor plus a separate
receptor (or receptors) for adenine nucleotides. An
important question remains regarding the specificity of
the nucleotide receptor. Since the receptor responds to
ADP and APCP, it is clear that it recognizes diphosphates.
However, the triphosphates ATP and ACPPP might cause their
responses after being converted to the corresponding diphosphates. Since the receptor clearly recognizes ADP, and
since it is unclear whether recognition extends to ATP, it
is probably best at present to refer to the receptor as an
ADP receptor. The receptor in taenia coli however is
obviously not the same as the ADP receptor in platelets,
because the latter has a very weak affinity for APCP (45).
It is possible that there is more than one nucleotide
receptor in taenia coli (46).

The guinea pig bladder obviously contains two different
purinoreceptors, since ATP and ADP contract this tissue,
while adenosine and AMP cause relaxations (47). In this case,
ATP is more potent than ADP, so the nucleotide receptor should
probably be classified as an ATP receptor. Several other
tissues are thought to have more than one purinoreceptor
(48-50).

Burnstock (16) has proposed that purinoreceptors be
divided into P_1 and P_2 classes, P_1 receptors being
adenosine receptors and P_2 receptors being receptors for
adenine nucleotides. This classification scheme has the
drawback of placing ADP and ATP receptors in the same P_2
class. In addition, there is no obvious advantage in
putting receptors for adenosine (a neutral compound) under
the same conceptual umbrella as receptors for ADP and ATP
(multivalent anions). To date, there is no evidence that
adenosine, ADP, and ATP receptors are related in the evolu-
tionary sense or in the sense of sharing similar post-receptor

mechanisms. For this reason, the author prefers to name each receptor after its preferred ligand, i.e., adenosine receptors, ADP receptors, ATP receptors.

There are several characteristics which are shared by nucleotide receptors. Responses mediated by true nucleotide receptors are not blocked by theophylline or adenosine deaminase, so these agents can be used to eliminate the possibility that responses to nucleotides might be mediated by breakdown of the nucleotides to adenosine. As mentioned above, many nucleotide receptors mediate responses which are opposite to the response to adenosine (Table IV). In addition, most nucleotide receptors respond to nucleotide analogs which are not subject to enzymatic hydrolysis, such as APCP, APCPP, and ACPPP. These analogs thus become powerful tools for the characterization of nucleotide receptors. In this context, it should be noted that β,γ-substituted nucleotides such as APPCP (β,γ-methylene-ATP) and APPNP (adenylylimidodiphosphate) are hydrolyzed by nucleotide triphosphate pyrophosphatase to AMP (42,51), and thus responses to β,γ-analogs do not necessarily imply the existence of a nucleotide receptor. Finally, several compounds which block nucleotide receptor-mediated responses are known. With two exceptions, these compounds appear to block postreceptor events and not the receptor itself. Several compounds such as quinidine, phentolamine, and 2,2'-pyridylisatogen tosylate block responses to ATP (16,52), but do not appear to have a high degree of specificity (52). Indomethacin and other cyclooxygenase inhibitors block responses to ATP in some tissues, implying that some of the responses to ATP are due to secondary release of prostaglandins or thromboxanes (46). Apamin, a peptide from bee venom which is a very potent blocker of calcium-dependent K^+ channels (53), blocks some of the responses to ATP (54). Two competitive antagonists of nucleotide receptors exist. As mentioned above, ATP appear to be competitive blockers at the platelet ADP receptor (37). In addition, the photoaffinity label aryl-azido aminopropionyl ATP irreversibly blocks the responses to ATP in some (54) but not all (55) tissues. Since ATP can prevent the irreversible blockade (54), the mode of action appears to be competitive. Arylazido aminopropionyl ATP has been used to support (56,57) or disprove (58) the hypothesis that ATP is a neurotransmitter in different tissues.

VIII.   FUTURE DEVELOPMENTS

In summary, then, good progress has been made on the adenosine receptor, and we are coming close to understanding adenosine's roles in vivo. In contrast, research on nucleo-tide receptors is still in its infancy. At this point we have little structure-activity data, few research tools such as competitive antagonists, no clear idea about how many receptors there are (a case can be made for at least a half-dozen types and subtypes of nucleotide receptors), and little evidence about possible roles of these receptors in vivo. The nucleotide receptor field thus should provide many opportun-ities for fundamental discoveries in the future.

ACKNOWLEDGEMENTS

The fibroblast experiments presented in this paper were done in the laboratory of Dr. Charles E. Spooner, the binding experiments were done in the laboratories of Dr. Solomon H. Snyder and Dr. John W. Daly, and the behavioral experiments were done by Jefferson Katims in the laboratory of Dr. Zoltan Annau. The author thanks Grace Wood and Gina Lu for skillful technical assistance and Connie Cook for intrepid word pro-cessing. The author also thanks the many chemists who provided compounds for these studies.

REFERENCES

1. Drury, A.N., and Szent-György, A., J. Physiol.
    London 68, 213 (1929).

2. Daly, J.W., J. Med. Chem. 25, 197 (1982).

3. Spielman, W.S. and Thompson, C. I., Am. J. Physiol.
    262, F423 (1982).

4. Sattin, A. and Rall, T.W., Mol. Pharmacol. 6, 13 (1970)

5. Bruns, R.F., Naunyn-Schmiedeberg's Arch. Pharmacol.,
    315, 5 (1980).

6. Kelly, L.A., Hall, M.S., and Butcher, R.W., J. Biol.
    Chem. 249, 5182 (1974).

7. Londos, C., Cooper, D.M.F., Schlegel, W., and Rodbell, M.,
    Proc. Natl. Acad. Sci. U.S.A. 75, 5362 (1978).

8. van Calker, D., Muller, M., and Hamprecht, B., J. Neuro-
    chem. 33, 999 (1979).

9. Schrader, J., Rubio, R., and Berne, R.M., J. Mol. Cell.
    Cardiol. 7, 427 (1975).

10. Olsson, R.A., Davis, C.J., Khouri, E.M., and Patterson,
    R.E., Circ. Res. 39, 93 (1976).

11. Bruns, R. F., Canad. J. Physiol. Pharmacol. 58, 673
    (1980).

12. Huang, M., Shimizu, H., and Daly, J.W., J. Med. Chem. 15,
    462 (1972).

13. Lai, T.F. and Marsh, R.E., Acta Crystallogr. B28, 1982
    (1972).

14. Premont, J., Guillon, G., and Bockaert, J., Biochem. Bio-
    phys. Res. Commun. 90, 513 (1979).

15. Londos, C. and Wolff, J., Proc. Natl. Acad. Sci. U.S.A.
    74, 5482 (1977).

16. Burnstock, G. and Brown, C.M., in "Purinergic Receptors"
    (G. Burnstock, ed.), p. 1. Chapman and Hall, London,
    1981.

17.   Bruns, R.F., Biochem. Pharmacol. 30, 325 (1981).

18.   Smellie, F.W., Davis, C.V., Daly, J.W., and Wells, J.N.,
      Life Sci. 24, 2475 (1979).

19.   Londos, C., Cooper, D.M.F., and Wolff, J., Proc. Natl.
      Acad. Sci. U.S.A. 77, 2551 (1980).

20.   Trost, T., and Stock, K., Naunyn-Schmiedeberg's Arch.
      Pharmacol. 299, 33 (1977).

21.   Bruns, R.F., Daly, J.W., and Snyder, S.H., Proc. Natl.
      Acad. Sci. U.S.A. 77, 5547 (1980).

22.   Smellie, F.W., Daly, J.W., Dunwiddie, T.V., and Hoffer,
      B.J., Life Sci. 25, 1739 (1979).

23.   Rodbell, M., Nature 284, 17 (1980).

24.   Schwabe, U., and Trost, T., Naunyn-Schmiedeberg's Arch.
      Pharmacol. 313, 179 (1980).

25.   Williams, M., and Risley, E., Proc. Natl. Acad. Sci.
      U.S.A. 77, 6892 (1980).

26.   Wu, P.H., Phillis, J.W., Balls, K., and Rinaldi, B.,
      Canad. J. Physiol. Pharmacol. 58, 576 (1980).

27.   Yeung, S.-M., and Green, R.D., Pharmacologist 23, 184
      (1981).

28.   Murphy, K.M.M. and Snyder, S.H., Life Sci. 28, 917
      (1980).

29.   Haulica, I., Ababei, L., Branisteanu, D., and
      Topoliceanu, F., J. Neurochem. 21, 1019 (1973).

30.   Vapaatalo, H., Onken, D., Neuvonen, P. J., and Westermann,
      E., Arzneim. Forsch. 25, 407 (1975).

31.   Daly, J.W., Bruns, R.F., and Snyder, S.H., Life Sci. 28,
      2083 (1981).

32.   Snyder, S.H., Katims, J.J., Annau, Z., Bruns, R.F., and
      Daly, J.W., Proc. Natl. Acad. Sci. U.S.A. 78, 3260 (1981).

33.   Beavo, J.A., Rogers, N.L., Crofford, O.B., Hardman, J.G.,
      Sutherland, E.W., and Newman, E.V., Mol. Pharmacol. 6,
      597 (1970).

34. Diamant, B., and Kruger, P.G., Acta. Physiol. Scand. 71, 291 (1967).

35. Haslam, R.J. and Cusack, N.J., in "Purinergic Receptors" (G. Burnstock, ed.) p. 221. Chapman and Hall, London, 1981.

36. Mills, D.C.B. and Smith, J.B., Biochem. J. 121, 185 (1971).

37. Cusack, N.J. and Hourani, S.M.O., Br. J. Pharmacol. 77, 329 (1982).

38. Cusack, N.J. and Hourani, S.M.O., Br. J. Pharmacol. 75, 397 (1982).

39. Macfarlane, D.E., Srivastava, P.C., and Mills, D.C.B., Thromb. Haemostas. 42, 185 (1979).

40. Marquardt, D.L., Parker, C.W., and Sullivan T.J., J. Immunol. 120, 871 (1978).

41. Satchell, D.G., and Maguire, M.H., J. Pharmacol. Exp. Ther. 195, 540 (1975).

42. Maguire, M.H. and Satchell, D.G., J. Pharmacol. Exp. Ther. 211, 626 (1979).

43. Brown, C.M. and Burnstock, G., Br. J. Pharmacol. 73, 617 (1981).

44. Satchell, D.G. and Maguire, M.H., Eur. J. Pharmacol. 81, 669 (1982).

45. Horak, H. and Barton, P.G., Biochim. Biophys. Acta. 373, 471 (1974).

46. Brown, C.M. and Burnstock, G., Eur. J. Pharmacol. 69, 81 (1981).

47. Burnstock, G., Dumsday, B., and Smythe, A., Br. J. Pharmacol. 44, 451 (1972).

48. Huizinga, J.D., Pielkenrood, J.M., and Den Hertog, A., Eur. J. Pharmacol. 74, 175 (1981).

49. Burnstock, G. and Megji, P., Br. J. Pharmacol. 73, 879 (1981).

50. Schwartzman, M., Pinkas, R., and Raz, A., Eur. J. Pharmacol. 74, 167 (1981).

51. Johnson, R. J., and Welden, J., Arch. Biochem. Biophys. 183, 216 (1977).

52. Maguire, M.H., and Satchell, D.G., in "Purinergic Receptors" (G. Burnstock, ed.) p. 47. Chapman and Hall, London, 1981.

53. Banks, B.E.C., Brown, C., Burgess, G.M., Burnstock, G., Claret, M., Cocks, T.M., Jenkinson, D.H., Nature 282, 415 (1979).

54. Hogaboom, G.K., O'Donnell, J.P., and Fedan, J.S., Science 208, 1273 (1980).

55. Frew, R., and Lundy, P.M., Life Sci. 30, 259 (1982).

56. Fedan, J.S., Hogaboom, G.K., O'Donnell, J.P., Colby, J., and Westfall, D.P., Eur. J. Pharmacol. 69, 41 (1981).

57. Theobald, R.J. Jr., J. Auton. Pharmacol. 3, 175 (1982).

58. Westfall, D.P., Hogaboom, G.K., Colby, J., O'Donnell, J.P., and Fedan, J.S., Proc. Natl. Acad. Sci. USA 79, 7041 (1982).

ANALOGS OF 2',5'-OLIGOADENYLATES: BIOLOGICAL

PROBES FOR THE ANTIVIRAL/ANTITUMOR STATE OF

MAMMALIAN CELLS[1]

Robert J. Suhadolnik, Yair Devash, and Paul Doetsch
Department of Biochemistry
Temple University School of Medicine
Philadelphia, Pennsylvania

Earl E. Henderson
Department of Microbiology and Immunology
Fels Research Institute
Temple University School of Medicine
Philadelphia, Pennsylvania

Joseph M. Wu
Department of Biochemistry
New York Medical College
Valhalla, New York

Wolfgang Pfleiderer and Ramamurthy Charubala
Fakultät Für Chemie
Universität Konstanz
Konstanz, Germany

I. INTRODUCTION

The secretion of interferon after viral infection enhances

antiviral resistance in other cells characterized by increased

[1]Supported by NSF research grant PCM-8111752 (RJS), NIH research
grants GM26134 (RJS), GM27210 (JMW), CA12227 (EEH), CA23999
(EEH) and U.S. Public Health Service training grant 5-T32
AM07162.

activity of certain enzymes, a protein kinase, an endoribonuclease and a (2'-5')oligoadenylate synthetase (1-6). The
biochemical function of the synthetase is to convert ATP into
oligonucleotides called (2'-5')pppA(pA)$_n$, which are
characterized by the 2',5'-phosphodiester bond. One of the
functions of these (2'-5')oligonucleotides is the activation
of the latent endoribonuclease (RNase L) which hydrolyzes RNA
(whose consequence is to inhibit protein synthesis). Because
of the proven therapeutic effects of interferon, interest has
developed in its use as an antiviral agent (7,8). However,
the use of interferon is limited due to scarcity, delivery to
target cells, antibody formation, and other adverse effects
(9,10). Because of these limitations, it may be possible to
bypass interferon by designing antiviral compounds with
specificity against virus-infected cells, but with minimum
cytotoxicity.

Our experience in the biosynthesis of the naturally
occurring nucleoside antibiotics and their use in prokaryotic
and eukaryotic systems (11,12) attracted us to the study of
structurally modified (2'-5')A$_n$[2] as biological probes in the

---

[2]Abbreviation: (2'-5')A$_n$, oligomer of adenylic acid with
(2'-5')phosphodiester linkages and a triphosphate at the
3'-end; core, 5'-dephosphorylated oligonucleotide.

study of the antiviral/antitumor state of mammalian cells. In
this chapter, we present investigations carried out in our
laboratories over the past few years concerning:

-enzymes of interferon action and requirements for

hydrolysis of RNA by the $(2'-5')A_n$-dependent nuclease

induced following virus infection

-regulation of the antiviral state in mammalian cells by

the intracellular accumulation of structurally modified

$(2'-5')A_n$

-inhibition of viral transformation and virus infection

of human lymphocytes by core $(2'-5')$oligonucleotides

in the absence of interferon

In view of the reported data on $(2'-5')A_n$, it seemed
logical that a minimal modification in the adenine moiety and
the ribosyl moiety would be of utmost use in the design of a
functional $(2'-5')A_n$ analog. This was accomplished by replacement of the 3′-hydroxyl group of the ribosyl moiety in
$(2'-5')A_n$ with a hydrogen atom and replacement of the 6-amino
group of the adenine in $(2'-5')A_n$ with a 6-hydroxyl group.
These structural changes resulted in the formation of
$(2'-5')3'dA_n$ and $(2'-5')I/A_n$ respectively (Fig. 1).

There are many ATP analogs that might replace ATP as a
substrate for $(2'-5')A_n$ synthetase in the formation of
$(2'-5')A_n$. Until 1965, there was no evidence for the formation of naturally occurring $(2'-5')$oligonucleotides. However,

experiments with cordycepin (3'-deoxyadenosine) in H. Ep. #1

cells indicated this might be a possibility (13). This

finding strongly suggested that 3'dATP might be a substrate

for the $(2'-5')A_n$ synthetase. If so, a $(2'-5')3'$-deoxy-

adenylate analog (lacking a 3'-hydroxyl group) might inhibit

translation, not be hydrolysed, nor be bound to 2',5'-phospho-

diesterase.

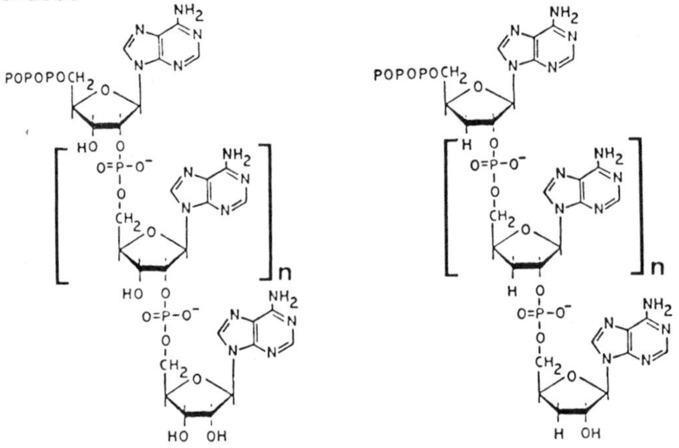

I   II

FIGURE 1.   Structures of the 5'-triphosphates of
$(2'-5')A_n$ (I) and $(2'-5')3'dA_n$ (II).

Such a $(2'-5')3'$-deoxyadenylate analog would be an excellent

$(2'-5')A_n$ probe for the study of substrate specificity of the

synthetase and inhibition of protein synthesis because once

synthesized, it would remain in the intracellular pool of the

cell.   Furthermore, we have reported that when the AMP moiety

of NAD[+] is replaced with 3'dAMP, there are marked changes in

DNA repair, poly(ADP-ribosylation) and formation of a produc-

tive ternary complex for dehydrogenase function (14,15).   In

addition, cordycepin has contributed greatly to our understanding of the formation of polyadenylate on the 3'-terminus
of mRNA and methylation on the 5'-terminus to form the "5'-cap"
(16-18, for review see 12).

We chose to modify the 6-amino group of $(2'-5')A_n$
because agents such as nitrous acid deaminate adenine and
cytosine-containing nucleosides and nucleotides in the cell.
This deamination has potential adverse effects on human health.
For example, deamination of dAMP to dIMP in DNA results in
AT-GC transition mutations. Furthermore, deamination of
$(2'-5')A$ to either $(2'-5')I_n$ or $(2'-5')I/A_n$ would allow us to
determing if base pairing of $(2'-5')$oligonucleotides with RNA
is required for RNA hydrolysis by the $(2'-5')A_n$-dependent
nuclease.

II. SYNTHESIS AND BIOCHEMICAL PROPERTIES OF $(2'-5')3'dA_n$

A. Enzymatic and Chemical Syntheses of $(2'-5')3'dA_n$ and Core $(2'-5')3'dA_n$.

When 3'dATP was incubated with either lysed rabbit
reticulocytes, L cell extracts, or HeLa cell extracts in
which the $(2'-5')A_n$ synthetase was bound to poly(rI)·poly(rC)-
agarose columns, $(2'-5')3'dA_n$ was formed (19). Product
formation was time-dependent. Following gradient elution
with DEAE cellulose, radioactivity was detected in fractions
with charge markers of -5 [$(2'-5')p_33'dA_2$ or $(2'-5')p_23'dA_3$],

$-6[(2'-5')p_33'dA_3$ or $(2'-5')p$ $3'dA_4]$ and -7 $[(2'-5')p_33'dA_4$ or $(2'-5')p_23'dA_5]$.

The replacement of ATP with 3'dATP as a substrate for $(2'-5')A_n$ synthetase has been reported from other laboratories. Lengyel and coworkers showed that $(2'-5')A_n$ synthetase from interferon-treated Ehrlich ascites tumor cells can link adenylate moieties to the 2'-hydroxyl of 3'dATP (20). While our studies on the enzymatic synthesis of $(2'-5')3'dA_n$ were in progress, Justesen et al. (21) reported that 3'dATP added to the 3'-end of $(2'-5')A_n$ following 3 hr incubations with rabbit reticulocyte $(2'-5')A_n$ synthetase. Minks et al. (22) reported that 3'dATP inhibits the HeLa cell $(2'-5')A_n$ synthetase in the presence of ATP. A tentative structure, $(2'-5')3'dA_3$, was assigned following enzymatic hydrolysis with BAP, SVPD I, and T2 RNase (19). Identification of the putative 5'-dephosphorylated core $(2'-5')3'dA_3$ was based on comparison with chemically synthesized core $(2'-5')3'dA_3$ (23). Alkaline hydrolysis (according to the procedure of Samanta et al. (20)) of the putative $[^{32}P](2'-5')p_33'dA_3$ or $[^{32}P](2'-5')p_23'dA_4$ isolated from DEAE cellulose did not produce any ^{32}P-labeled pppAp or 2',3'-AMP.

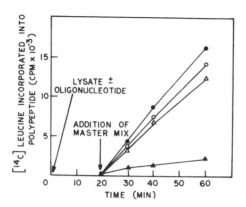

FIGURE 2. Inhibition of translation by (2'-5')A$_3$ and
(2'-5')3'dA$_n$. Inhibition of translation with (2'-5')A$_3$
(100 nM, Δ) and 100 nM plus periodate (o) were compared
with that of (2'-5')3'dA$_n$ (100 nM) or 100 nM plus
periodate ($\blacktriangle$) and control ($\bullet$).

B. Inhibition of Protein Synthesis by (2'-5')3'dA$_n$.

The inhibition of protein synthesis by (2'-5')3'dA$_n$ was

compared with the inhibition by (2'-5')A$_3$ (trimer 5'-triphos-

phate) (Fig. 2)(19). At equimolar concentrations, (2'-5')3'dA$_n$

(n=3 or 4, charge -6 or -7) is a better inhibitor of transla-

tion than (2'-5')A$_3$ (Fig. 2) and (2'-5')A$_4$ (tetramer).

C. Metabolic Stability of (2'-5')3'dA$_n$.

(2'-5')3'dA$_n$, when added to extracts of HeLa cells,

L cells, or C85-5C lymphoblasts, is not a substrate for

2',5'-phosphodiesterase (Fig. 3). Furthermore, (2'-5')3'dA$_n$

does not appear to compete for hydrolysis of (2'-5')A$_n$ by

2',5'phosphodiesterase nor is it bound to the 2',5'-phospho-
diesterase.

   D.   Intracellular Accumulation of $(2'-5')3'dA_n$.

Although considerable information has been obtained in
in vitro studies with $(2'-5')A_n$ with respect to protein
synthesis, activation of $(2'-5')A_n$-dependent nuclease, and
binding protein, little is known concerning the regulation

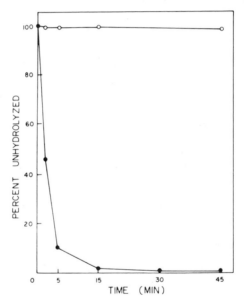

FIGURE 3.   Stability of $(2'-5')A_n$ (●) and $(2'-5')3'dA_n$
(o) in HeLa cell extracts (19).

of the antiviral state.   Thang and Kerr and their coworkers
(24,25) have reported on changes in binding protein following
virus infection of mammalian cells and the changes in
$(2'-5')A_n$ synthetase activity.   Although Baglioni and

coworkers (26) reported on the intracellular accumulation of
(2′-5′)A_n in interferon-poly(rI)·poly(rC) treated HeLa cells,
the fact that (2′-5′)A_n is rapidly metabolized does not permit
long term in vivo studies with respect to the enzymes important
in the antiviral state. Therefore, it was of interest to
determine whether 3′-deoxyadenosine could be taken up by mammalian cells and be converted to (2′-5′)3′dA_n in measurable
quantities. We have demonstrated that the addition of 2 µM
deoxycoformycin and 80 µM cordycepin to HeLa cells results in
an increase in the intracellular concentration of 3′dATP to
1.7 mM, while ATP decreased to zero (Fig. 4). This decrease
in ATP might be explained by the report of Bagnara and
Hershfield (27).

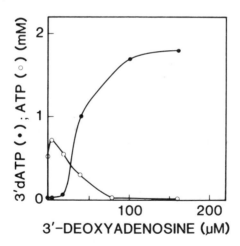

FIGURE 4.   Changes in intracellular concentrations of
ATP (o) and 3′dATP (•) in HeLa cells treated with deoxycoformycin (2 µM) and cordycepin (80 µM).

When [8-^{3}H]3'-deoxyadenosine (80 μM) was added to HeLa

cells treated with human fibroblast interferon (IFN-β, 200

units/ml), poly(rI)·poly(rC) (200 μg/ml) and deoxycoformycin

(2 μM), (2'-5')3'dA$_n$ was isolated by DEAE cellulose chroma-

tography. Dialysis of the 350 mM KCl eluant followed by

enzymatic hydrolysis (BAP, SVPD, and T2 RNase) showed that a

radioactive (2'-5')-oligonucleotide was displaced from the

DEAE column. All of the radioactivity in the product formed

following BAP digestion had the same R_f as authentic

chemically synthesized core (2'-5')3'dA$_3$. Because core

(2'-5')A$_3$ and A$_4$ have the same R_f in isobutyric acid:ammonia:

water (66:1:33, v/v/v) (solvent A), core (2'-5')3'dA$_4$ could

have cochromatographed with core (2'-5')3'dA$_3$. [^{3}H]3'dAMP,

but not [^{3}H]AMP, was isolated following SVPD and tlc in

solvent A. Protein synthesis was inhibited by (2'-5')3'dA$_n$

in rabbit reticulocyte lysate cell-free systems. The antiviral

state, based on plaque forming units per ml, decreased 600-

fold from control to treated HeLa cells. Treated cells also

showed induced (2'-5')A$_n$ synthetase and were viable as

measured by trypan blue exclusion.

The intracellular accumulation of the putative

(2'-5')3'dA$_n$ to a concentration of 32 nM compares with previous

reports on the intracellular accumulation of (2'-5')A$_n$ to

600 nM in IFN-β/dsRNA-treated HeLa cells (26) and to 2-200 nM

in interferon-treated virus-infected cells (28,29). The lack of hydrolysis of $(2'-5')3'dA_n$ will allow the study of effects of prolonged exposure of mammalian cells to this or other $(2'-5')A_n$ analogs without the use of calcium coprecipitation or permeabilization techniques. The intracellular accumulation of $(2'-5')A_n$ analogs might offer a promising approach to determine if the mode of action of $(2'-5')$-oligonuclgotides involves hydrolysis of mRNA and 5'-cap synthesis (30). $(2'-5')A_n$ analogs can also be used in the study of the regulation of $(2'-5')A_n$ synthetase and $(2'-5')A_n$-dependent nuclease as reported by Thang and coworkers (25).

III. SYNTHESIS AND BIOCHEMICAL PROPERTIES OF $(2'-5')I/A_4$

A. Chemical Synthesis of $(2'-5')I_n$ or $(2'-5')I/A_n$.

ITP was not a substrate for $(2'-5')A_n$ synthetase in lysed rabbit reticulocytes (31); therefore, synthesis of $(2'-5'I/A_3{}^3$ and $(2'-5')I/A_4$ was accomplished by chemical deamination of $(2'-5')A_3$ and $(2'-5')A_4$ with nitrous acid (32). The structure of the putative $(2'-5')I/A_3$ and $(2'-5')I/A_4$ was established by enzyme and acid hydrolyses and PEI thin layer chromatography. Following deamination, the molar ratio of IMP:AMP in $(2'-5')I/A_{3,4}$ was 0.85:0.15.

---

[3]$(2'-5')I/A_3$ and $(2'-5')I/A_4$ = deaminated trimer and tetramer

However, the distribution of the IMP and AMP in the nucleotide
has not been established.

B. Inhibition of Protein Synthesis by $(2'-5')I/A_3$ and
$(2'-5')I/A_4$.

Lysed rabbit reticulocytes preincubated 10 minutes with
$(2'-5')A_4$, $(2'-5')I/A_4$ or $(2'-5')3'dA_n$ at 1, 10 and 30 nM
showed inhibition of translation (33)(Fig. 5). Whereas 30 nM
$(2'-5')I/A_3$ and $(2'-5')A_3$ show no inhibition of protein
synthesis, 30 nM $(2'-5')3'dA_n$ inhibits protein synthesis 72%
after 60 minutes. $(2'-5')3'dA_n$ (◆) is as good an inhibitor
as is tetramer $(2'-5')A_4$ (◇) at 30 nM following 60 minute
incubation (Figure 5). Assays with $(2'-5')I/A_4$ showed
inhibition of protein synthesis similar to that observed for
$(2'-5')A_4$. SVPD treatment of $(2'-5')A_4$ and $(2'-5')I/A_4$
showed no inhibition of protein synthesis.

C. Activation of $(2'-5')A_n$-dependent Nuclease.

To determine if $(2'-5')I_4$ and $(2'-5')3'dA_n$ inhibit
protein synthesis by activation of the $(2'-5')A_n$-dependent
nuclease, vesicular stomatitis, vaccinia and reovirus mRNAs
were incubated with partially purified $(2'-5')A_n$-dependent
nuclease. All mRNAs were hydrolyzed by $(2'-5')A_4$, $(2'-5')I_4$,
and $(2'-5 ')3'dA_n$ at 10 and 30 nM, while $(2'-5')A_n$ cleaves
VSV mRNA 42 and 64% (33).

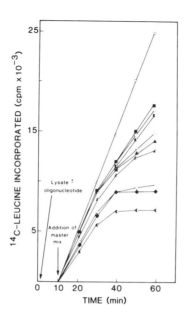

FIGURE 5. Inhibition of protein synthesis by (2'-5')A4, (2'-5')I/A4, and (2'-5')3'dAn. The inhibition of translation by (2'-5')A4, (2'-5')I/A4, and (2'-5')3'dA at 1 nM (□,■,◨), 10 nM (△,▲,◮) and 30 nM (◇,◆,◈) were compared with control assays (o); (o) also represents the effect on translation by chemically synthesized core (2'-5')3'dA3, core (2'-5')I3 core (2'-5')I4 and (2'-5')/A3 at 30 nM. Open symbols (2'-5')A4; closed symbols, (2'-5')I/A4; half-closed symbols, (2'-5')3'dAn.

D. Competition of (2'-5')A4, (2'-5')I/A4 and
(2'-5')3'dAn for (2'-5')A4[32P]pCp.

(2'-5')A4 [32P]pCp is used as a (2'-5')An probe for

binding protein activity because it is not hydrolyzed in cell

extracts. There was a rapid competition of (2'-5')A4,

(2'-5')I/A4 and (2'-5')3'dAn for (2'-5')A4[32P]pCp with

either lysed rabbit reticulocytes or partially purified

(2'-5')An-dependent nuclease (Fig. 6)(33). However,

enzymatically synthesized (2'-5')3'dAn (350 mM KCl fraction,

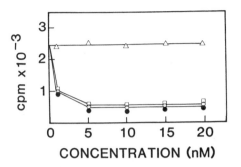

CONCENTRATION (nM)

FIGURE 6. Binding of $(2'-5')A_4[^{32}P]$ to partially purified $(2'-5')A_n$-dependent nuclease. The amount of
$(2'-5')A_4[^{32}P]pCp$ competed off nitrocellulose discs by
increasing concentrations of $(2'-5')A_3$ or $(2'-5')3'dA_n$
purified on DEAE-7 M urea ($\bullet$), $(2'-5')A_4$ (o), $(2'-5')I/A_4$
($\square$); $(2'-5')3'dA_n$ (350 mM KC1 fraction)($\triangle$, 0-1 nM).

$\triangle$, 0-1 nM) was not of sufficient concentration to displace

the pCp.

Because of changes in competition of $(2'-5')3'dA_n$ for

the pCp analog, it was essential to determine if $(2'-5')3'dA_n$

does in fact bind to the binding protein. Experimentally,

$[^3H](2'-5')3'dA_n$ (20 Ci/mmole) was indeed bound to partially-

purified $(2'-5')A_n$-dependent nuclease and was displaced by

increasing concentrations of unlabeled $(2'-5')3'dA_4$ (33).

The inhibition of protein synthesis by $(2'-5')I/A_n$

raises interesting questions concerning the activation of the

$(2'-5')A_n$-dependent nuclease. If adenylate residues from

$(2'-5')A_n$ are required for base pairing with UMP residues in

RNA for the hydrolysis of RNA as has been suggested (34,35),

then the deamination of $(2'-5')A_n$ to $(2'-5')I/A_n$ may change

the base pairing. These studies are in progress. Finally,

if $(2'-5')I/A_4$ hydrolyzes only poly(U), as shown for

(2′-5′)A_n, then the activation of the (2′-5′)A_n-dependent
nuclease by (2′-5′)oligonucleotides might require other
changes in the enzyme. By making structural changes in
either the aglycone or ribosyl moiety of (2′-5′)A_n, we could
have biochemical probes that would enable us to determine if
the oligonucleotides activate the nuclease by base pairing or
if another mechanism is involved.

IV. ANTIVIRAL PROPERTIES OF CORE (2′-5′)3′dA_n AND (2′-5′)I_3

A.   Inhibition of EBV-Induced Transformation of Human
                    Lymphocytes.

     Epstein-Barr virus (EBV) is a lymphotrophic, double-
stranded DNA virus of the herpes virus group. EBV infection
is the cause of infectious mononucleosis and is implicated in
the etiology of three human cancers (36). We chose EBV-
infected lymphocytes to study the effect of (2′-5′)oligo-
nucleotides on virus-induced transformation. The 5′-dephos-
phorylated (2′-5′)oligonucleotides (cores) were known to be
taken up by concanavalin A-stimulated lymphocytes (37). We
designed experiments to determine if chemically synthesized
core (2′-5′)A_n, (2′-5′)I_3, and (2′-5′)3′dA_3 could prevent the
transformation of EBV-infected lymphocytes in the absence of
human interferon. Enzymatically synthesized core (2′-5′)3′dA_n
gave the same results as chemically synthesized core
(2′-5′)3′dA_3.

1. EBV-induced DNA Synthesis. Infection of lymphocytes by EBV is known to induce a stimulation of lymphocyte (host cell) DNA synthesis. EBV-infected lymphocytes showed a 2.4-fold increase in host cell DNA synthesis compared to uninfected cells (Table I)(38). Core $(2'-5')A_3$, $(2'-5')A_4$ and $(2'-5')3'dA_3$ showed a dose-dependent inhibition of EBV-induced DNA synthesis. Core $(3'-5')A_3$ had no effect. IFN-β reduced DNA synthesis in EBV-infected cultures to 60% that of control infected cells and increased DNA synthesis in uninfected cells.

2. EBV-induced Morphological Transformation. The transformed centers assay was also used to determine the effect of interferon and core $(2'-5')$oligonucleotides on EBV-induced transformation of lymphocytes. Transformed colonies were

TABLE I.   Effect of IFN-β and Core (2'-5')Oligonucleotides on
           Spontaneous and EBV-induced DNA Synthesis in
           Lymphocytes (38)

Treatment	[³H] Thymidine incorporation			
	Infected		Uninfected	
	cpm	ratio[a]	cpm	ratio
None (control)	17,450	1.00	7,390	1.00
IFN-β				
10 units/ml	15,000	0.86	9,090	1.22
100 units/ml	11,990	0.69	10,460	1.41
250 units/ml	10,420	0.60	16,550	2.23
Core (2'-5')A_3				
25 μM	7,950	0.45	5,050	0.68
50 μM	4,013	0.23	4,411	0.59
200 μM	1,770	0.10	1,870	0.25
Core (2'-5')A_4				
5 μM	10,580	0.60	6,190	0.84
25 μM	6,460	0.37	3,560	0.48
100 μM	740	0.04	1,560	0.21
Core (2'-5')3'dA_3				
5 μM	17,790	1.02	7,420	1.00
100 μM	13,380	0.77	7,760	1.03
200 μM	8,150	0.47	11,580	1.56

[a]ratio of treated to nontreated cells.

measured 4-6 weeks after infection with EBV and simultaneous

treatment with core oligonucleotides or interferon (38). Core

(2'-5')3'dA$_3$ (75-100 µM) inhibited the formation of transformed colonies. This inhibition was dose-dependent. In
parallel studies, human leukocyte interferon (IFN-α) was more
effective than human fibroblast interferon (IFN-β) in
preventing transformation (38). Core (2'-5')3'dA$_3$ and core
(2'-5')A$_4$ also inhibited the appearance of transformed
colonies.

Because ara-IMP is known to have weak antiviral properties
and because we have shown the antiviral/antitumor properties
of core (2'-5')3'dA$_3$, we next investigated the activity of
core (2'-5')I$_3$ with EBV-infected lymphocytes.

The limiting dilutions assay was used to determine the
effect of interferon and chemically synthesized core (2'-5')I$_3$
on EBV-induced transformation of peripheral blood lymphocytes.
IFN-α (500 units/ml), core (2'-5')A$_3$ (100 µM) and core
(2'-5')I$_3$ (50, 100, 200 µM) inhibited the appearance of transformed colonies. In two experiments, chemically synthesized
(2'-5')I$_3$ inhibited transformation in a dose-dependent manner.
However, a third experiment showed no inhibition of transformation. Further experimentation is necessary to resolve these
differences. Core (2'-5')I$_4$ did not inhibit transformation
of EBV-infected lymphocytes (39).

    3. <u>Lymphoblastoid Cell Survival</u>. We have tested the
possibility that the inhibition of lymphoblastoid transformation was due to cytotoxicity (38). Neither core

(2'-5')3'dA$_3$ nor core (2'-5')I$_3$ was cytotoxic at concentrations up to 300 μM. However, core (2'-5')A$_3$ (150-300 μM) and core (2'-5')A$_4$ (150-300 μM) were toxic to both Raji and BJAB lymphoblasts.

4. Epstein-Barr Virus Associated Nuclear Antigen (EBNA) Induction. To determine the mechanism by which core (2'-5')3'dA$_3$ inhibits EBV-induced transformation, we have investigated the inhibition of Epstein-Barr virus associated nuclear antigen (EBNA) formation (40). This study was accomplished by staining cells with fluorescein isothiocyanate conjugated goat antihuman complement. Areas of fluorescence associated with nuclei were scored for the presence of EBNA (Table II). The data suggest that the antiviral mechanism of (2'-5')3'dA$_3$ is through the inhibition of EBV early functions, possibly through transcription or translation of EBNA.

TABLE II. EBNA Appearance in Human Adult Lymphocytes
            Inoculated with Epstein-Barr Virus (40)

Treatment	Percentage of EBNA-positive cells		
	Expt. 1	Expt. 2	Expt. 3
None (control)	27	2	4
IFN-α			
500 units/ml	0	0	0
Core (2'-5')A$_3$			
300 μM	2	-	-
150 μM	0	1	-
Core (2'-5')A$_4$			
300 μM	0	-	-
150 μM	2	0	-
Core (2'-5')3'dA$_3$			
300 μM	0	-	-
150 μM	2	0	-
EBV-transformed	>90	-	>80
Uninfected control	0	<0.3	<0.1
MOLT cells (EBV-negative control)	0	-	-

5. Uptake of Core [^{32}P](2'-5')A$_n$ and Core [^{32}P]-
(2'-5')3'dA$_n$ by Lymphoblasts. Uptake of core [^{32}P](2'-5')A$_3$ and
[^{32}P](2'-5')3'dA$_n$ into the cytoplasm of lymphoblasts was

established by a modification of Knight et al. (41). After

12 hours of treatment with core $[^{32}P](2'-5')A_n$ the only

^{32}P-labeled compounds detected in C85-5C lymphoblast cell

extracts were ATP, ADP, AMP, and inorganic phosphate. After

6 hours incubation of core $[^{32}P](2'-5')3'dA_n$, the extract was

chromatographed on a cellulose tlc plate (solvent: isobutyric

acid/ammonium hydroxide/water, 66/1/33, v/v/v). Visualization

of the autoradiogram indicated that there was no hydrolysis

of core $[^{32}P](2'-5')3'dA_n$. However, when this lane was cut

into 1 cm^2 pieces and assayed for ^{32}P, 16% was detected as

inorganic phosphate (R_f 0.29), 4.6% as 3'dAMP (R_f 0.65) and

79% as core $[^{32}P](2'-5')3'dA_n$ (R_f 0.79). No 5'-triphosphate

of $(2'-5')3'dA_n$ was detected (R_f 0.36). After 6 hours,

96.5% of core $[^{32}P](2'-5')3'dA_3$ remained in HUCL medium.

B. Inhibition of Virus Infection by Core $(2'-5')3'dA_3$.

Because of the inhibition of transformation of EBV-

infected lymphocytes we observed with core $(2'-5')3'dA_n$, we

sought to determine if core $(2'-5')3'dA_3$ inhibits other DNA

viruses. We employed an infected centers assay utilizing

lymphocytes to determine the ability of (2'-5')oligonucleotides

to inhibit HSV-1 After incubation for 72 hr with or without

(2'-5')oligonucleotides, lymphocytes exposed to HSV-1 were

plated onto human foreskin fibroblast monolayers (Table III).

Lymphocytes not treated with IPN-β or (2'-5')oligonucleotides developed as infected centers. The fibroblast monolayers

onto which IFN-β treated lymphocytes were plated developed

significantly less cytopathic effects and then only when

higher infected cell concentrations (6.7 x 10^5 infected

cells) were plated onto the fibroblast monolayers. Monolayers

exposed to uninfected fibroblasts did not develop cytopathic

effects (cpe). Core (2'-5')A_3 and core (2'-5')A_4 inhibited

the development of cpe, with significant cpe observed only

with fibroblasts treated with 2 x 10^5 cells. Core

(2'-5')3'dA_3 was less effective in inhibiting the development

of cpe compared to IFN-β, core (2'-5')A_3 and core (2'-5')A_4

(42).

IV. CONCLUSIONS

This chapter describes studies performed in our

laboratories on the development of structurally modified

(2'-5')oligoadenylates. Our purpose was to examine require-

ments for the development of the antiviral/antitumor state in

mammalian cells. We sought to modify (2'-5')A_n to a minimal

extent to produce functional (2'-5')A_n analogs with differing

biological activities.

TABLE III. Effect of IFN-β and (2'-5')Oligonucleotides on
Development of Infected Centers from HSV-1
Infected Lymphocytes as Determine by Cytopathic
Effects (cpe) (42)

Treatment	cpe at given number of HSV-exposed lymphocytes per cell		
	2×10^5	6.7×10^4	2.2×10^4
None (control)	+++	++	+
IFN-β			
25 units/ml	++	+	0
100 units/ml	+++	0	0
250 units/ml	+	0	0
500 units/ml	0	0	0
Core (2'-5')A$_3$			
10 μM	+++	++	0
50 μM	++	++	0
75 μM	++	0	0
150 μM	+	0	0
Core (2'-5')A$_4$			
10 μM	++	+	0
50 μM	+	0	0
75 μM	++	0	0
150 μM	+	0	0
Core (2'-5')3'dA$_3$			
10 μM	+++	++	+
50 μM	++	++	+
75 μM	++	+	0
150 μM	-	-	0
Core (3'-5')A$_3$			
300 μM	+++	0	0

+++ = complete destruction of the monolayer
++ = multiple foci of cpe
+   = one focus of cpe
0   = no monolayer disturbance

Structural modifications were made in both the adenine and ribose of $(2'-5')A_n$. The 6-amino group of adenine or $(2'-5')A_4$ was partially deaminated to the 6-hydroxyl group to form $(2'-5')I/A_n$. The ribosyl moiety of $(2'-5')A_n$ was modified by replacing the 3'-hydroxyl group with a hydrogen atom to form $(2'-5')3'dA_n$.

The deamination of $(2'-5')A_n$ is of interest for several reasons. Deaminated analogs of nucleosides and nucleotides are known to have antiviral activity. For example, the antiviral activity against DNA viruses in cell culture following deamination of ara-A and ara-AMP to their corresponding inosine derivatives has been reported (43,44). Second, Lengyel and coworkers (34) have shown that activation of the $(2'-5')A_n$-dependent nuclease (RNase L) with $(2'-5')A_n$ will hydrolyze poly(U), but not poly(A), poly(G), or poly(C). Kerr and coworkers (35) have reported that the nuclease cleaves predominantly at the 3'-side of UU, UG and UA residues in RNA. Implicit in these observations is that the adenylate of $(2'-5')A_n$ base pairs with the uridylate of UMP in RNA in order to cleave RNA. If base pairing is essential for RNA cleavage, then deamination of $(2'-5')A_n$ could markedly affect the activation of the $(2'-5')A_n$-dependent nuclease. This is based on the fact that the putative $(2'-5')I/A_n$ activated the nuclease as does $(2'-5')A_n$. If base pairing is essential for activation of the nuclease, then $(2'-5')I/A_n$ base pairing with RNA might differ or not occur at all. Another possible

explanation for the activation of the nuclease by $(2'-5')I/A_n$

is that A-to-U base pairing is not essential for activation.

Perhaps $(2'-5')A_n$ causes a conformational change in the

nuclease. Another reason for deamination of $(2'-5')A_n$ was

that modification of purines and pyrimidines in RNA and DNA

is known to alter the interaction with protein and/or other

nucleotides. For example, the deamination of the adenine

residues of dAMP in DNA results in AT-GC transition mutations

(45). This deamination can occur nonenzymatically either by

nitrous acid or by heating at pH 7.6 (31,45). It is possible

that the deamination of $(2'-5')A_n$ could result in the in vivo

formation of either $(2'-5')I_n$ or a combination of $(2'-5')I/A_n$.

Such nucleotides could markedly affect the enzymes required

for interferon action and subsequently interfere with the

development of the antiviral/antitumor state of the mammalian

cell. If (2'-5')oligonucleotides containing random mixtures

of adenylate and inosinate exist in the cell due to deamination,

they might play a critical role in the resistance of the

mammalian cell to various viral infections. The cell may

contain many combinations of $(2'-5')I/A_n$, each of which is

the specific defense against a single invading virus.

    Cordycepin (3'-deoxyadenosine) was the first naturally

occurring nucleoside antibiotic isolated (46); it is a

cytostatic analog of adenosine (11). 3'dATP, although a

structural isomer of 2'dATP, is a functional analog of ATP.
3'dATP was selected for these studies because it had been
reported to form the (2'-5')phosphodiester bond in H. Ep. #1
cells (13), cause marked changes in DNA repair and dehydrogenase
activity through the formation of 3'dNAD$^+$ (14,15), and lacks
a 3'-hydroxyl group (thereby making it likely to be resistant
to hydrolysis by the 2',5'-phosphodiesterase present in cells).

We have been fortunate in that (2'-5')3'dA$_n$ has provided
us with considerable information with respect to the antiviral/
antitumor state of mammalian cells. Our results raise several
exciting possibilities for future studies. First, in relation
to the (2'-5')A$_n$-associated enzymes that are important in the
development of the antiviral state, we observed that
(2'-5')3'dA$_n$ inhibits protein synthesis, activates the
(2'-5')A$_n$-dependent nuclease which subsequently degrades
mRNA. [^{3}H](2'-5')3'dA can be displaced from the nuclease by
unlabeled (2'-5')3'dA$_n$. These results reveal that (2'-5')3'dA$_n$
can bind to the endonuclease. However, in some enzymatic
syntheses, (2'-5')3'dA$_n$ did not compete off the pCp analog.
These findings are currently under further investigation.
Our knowledge that (2'-5')3'dA$_n$ is not degraded by the
(2'-5')phosphodiesterase permitted the study of the intracellular accumulation of (2'-5')3'dA$_n$ in interferon/double
stranded RNA/deoxycoformycin/cordycepin treated HeLa cells.
The accumulation of a stable (2'-5')A$_n$ analog will permit the

study of its long term effects on the $(2'-5')A_n$-dependent
enzymes, RNA, DNA and protein synthesis in cells without
having to use the calcium coprecipitation technique, permeabilization, or interferon treatment. Based on our observation
that $(2'-5')3'dA_n$ was stable in lymphocytes as determined by
the uptake of $[^{32}P](2'-5')3'dA_3$ plus the report that core
$(2'-5')A_n$ was taken up by concanavalin A stimulated lymphocytes
(37), we completed experiments which showed that core $(2'-5')A_n$
and either enzymatically or chemically synthesized $(2'-5')3'dA_3$
prevent the transformation of EBV-infected lymphocytes.
Preliminary experiments show that chemically synthesized core
$(2'-5')A_n$ and $(2'-5')3'dA_3$ weakly inhibit HSV-1 infection of
lymphocytes (42). Another encouraging aspect of the core
$(2'-5')3'dA_3$ is that while the naturally occurring $(2'-5')A_n$
is cytotoxic to uninfected lymphocytes and proliferating
lymphoblasts, the cordycepin analog is not cytotoxic. It is
our hope that the inhibition of transformation of virus infection of mammalian cells by $(2'-5')A_n$ analogs will result in
therapeutic application. Preliminary experiments suggest
that core $[^{32}P](2'-5')3'dA_n$ might be taken up by uninfected
lymphocytes and lymphoblasts. Intact core $[^{32}P](2'-5')A_n$ was
detected in TCA extracts, but 5'-phosphorylation was not
detected. We can not state whether the mechanism of inhibition
of transformation is a core-membrane or core-intracellular
process. We are currently studying the metabolic fate of

radioactive core $(2'-5')3'dA_3$ in lymphocytes.

One of the difficulties remaining with core $(2'-5')A_n$ analogs is their ability to be transported and taken up by target cells in the intact animal. This biological problem might be overcome by the covalent addition of lipophilic groups (fatty acids) to either the C-5' hydroxyl or the 6-amino group of the $(2'-5')$oligonucleotides. Precedents for these changes are the reports by Baker and coworkers (47,48) showing that 5'-O-valeryl-ara-A was more toxic than ara-A. Furthermore, Revel and coworkers (49) showed that the butylated trimer with a free 5'-hydroxyl group inhibited DNA synthesis 60%, whereas the butylated or acetylated trimer with 5'-hydroxyl blocked did not inhibit DNA synthesis. Therefore, the 3'-hydroxyl of the trimer can be blocked and still show inhibitory activity.

These studies raise the exciting possibility that analogs of $(2'-5')A_n$ may be useful in the treatment of viral infections and cancers either alone or in combination with interferon or chemotherapeutic agents.

ACKNOWLEDGMENTS

The authors wish to acknowledge the excellent contributions of Y. Sawada and N. L. Reichenbach.

REFERENCES

1. Lengyel, P., Annu. Rev. Biochem. 51:251 (1982).

2. Pestka, S., ed., "Interferons", Methods in Enzymology,
    Vol. 78, Part A, Academic Press, New York, 1981.

3. Pestka, S., ed., "Interferons", Methods in Enzymology,
    Vol. 79, Part B, Academic Press, New York, 1981.

4. Baglioni, C., Cell 17:255 (1979).

5. Stewart, W. E. II, "The Interferon System", Springer,
    New York, 1979.

6. Hovanessian, A. G., Brown, R. E., and Kerr, I. M., Nature
    268:537 (1977).

7. Strander, H., Blut 35:277 (1977).

8. Marx, J. L., Science 204:1183 (1979).

9. Vallbracht, A., Treuner, J., Flehming, B., Joester, K.-
    E., and Niethammer, D., Nature 289:496 (1981).

10. Cantell, K., Endeavour 2:27 (1978).

11. Suhadolnik, R. J., "Nucleoside Antibiotics", John Wiley
    and Sons, Inc., New York, 1970.

12. Suhadolnik, R. J., "Nucleosides as Biological Probes",
    John Wiley and Sons, Inc., New York, 1979.

13. Cory, J. G., Suhadolnik, R. J., Resnick, B., and Rich,
    M. A., Biochim. Biophys. Acta 103:646 (1965).

14. Suhadolnik, R. J., Lennon, M. B., Uematsu, T., Monahan,
    J. E., and Baur, R., J. Biol. Chem. 252:4125 (1977).

15. Suhadolnik, R. J., Baur, R., Lichtenwalner, D. M., Uematsu, T., Roberts, J. H., Sudhakar, S., and Smulson, M., J. Biol. Chem. 252:4134 (1977).

16. Glazer, R. I., and Peale, A. L., Biochem. Biophys. Res. Commun. 81:521 (1978).

17. Kredich, N. M., J. Biol. Chem. 255:7380 (1980).

18. Sawicki, S. G., Jelinek, W., and Darnel, J. E., J. Molec. Biol. 113:219 (1977).

19. Doetsch, P., Wu, J. M., Sawada, Y., and Suhadolnik, R. J., Nature 291:355 (1981).

20. Samanta, H., Dougherty, J. P., and Lengyel, P., J. Biol. Chem. 255:9807 (1980).

21. Justesen, J., Ferbus, D., and Thang, M. W., Proc. Natl. Acad. Sci. U.S.A. 77:4618 (1980).

22. Minks, M. A., Benvin, S., and Baglioni, C., J. Biol. Chem. 255:5031 (1980).

23. Charubala, R., and Pfleiderer, W., Tetrahedron Lett. 21:4077 (1980).

24. Lab, M., Thang, M. N., Soteriadou, K., Koehren, F., and Justesen, J., Biochem. Biophys. Res. Commun. 105:412 (1982).

25. Cayley, P. J., Knight, M., and Kerr, I. M., Biochem. Biophys. Res. Commun. 104:376 (1982).

26. Nilsen, T. W., Maroney, P. A., and Baglioni, C., J. Biol. Chem. 256:7806 (1981).

27. Bagnara, A. S., and Hershfield, M. S., Proc. Natl. Acad. Sci. U.S.A. 79: 2673 (1982).

28. Williams, B. R. G., Golgher, R. R., Brown, R. E., Gilbert, C. S., and Kerr, I. M., Nature 282:582 (1979).

29. Nilsen, T. W., Maroney, P. A., and Baglioni, C., J. Virology 42:1039 (1982).

30. Goswami, B. B., Crea, R., Van Boom, J. H., and Sharma, O. K., J. Biol. Chem. 257:6867 (1982).

31. Suhadolnik, R. J., Doetsch, P. W., and Reichenbach, N. L., Fed. Proc. 41, Abstract 6927 (1982).

32. Schuster, V. H., Z. Naturforschung 15b:298 (1960).

33. Devash, Y., Doetsch, P. W., Wu, J. M., Reichenbach, N. L., Pfleiderer, W., Charubala, R., and Suhadolnik, R. J., manuscript in preparation.

34. Floyd-Smith, G., Slattery, E., and Lengyel, P., Science 212:1030 (1981).

35. Wreschner, D. H., McCauley, J. W., Skehel, J. J., and Kerr, I. M., Nature 289:414 (1981).

36. Epstein, M. A., and Achong, B. G., eds., "The Epstein Barr Virus", Springer, New York, 1979.

37. Kimchi, A., Shure, H., and Revel, M., Nature 282:849 (1979).

38. Doetsch, P. W., Suhadolnik, R. J., Sawada, Y., Mosca, J. D., Flick, M. B., Reichenbach, N. L., Dang, A. Q., Wu, J. M., Charubala, R., Pfleiderer, W., and Henderson, E. E., Proc. Natl. Acad. Sci. U.S.A. 78:6699 (1981).

39. Henderson, E. E., Devash, Y., and Suhadolnik, R. J., unpublished results.

40. Henderson, E. E., Doetsch, P. W., Charubala, R., Pfleiderer, W., and Suhadolnik, R. J., Virology 122:198 (1982).

41. Knight, M., Cayley, P. J., Silverman, R. H., Wreschner, D. H., Gilbert, C. S., Brown, R. E., and Kerr, I. M., Nature 288:189 (1980).

42. Henderson, E. E., Doetsch, P. W., Flick, M. B., and Suhadolnik, R. J., manuscript in preparation.

43. Reinke, C. M., Drach, J. C., Shipman, C., Jr., and Weissbach, A., "Oncogenesis and Herpes Viruses", III, Lyon, France, IARC Press, p. 999 (1978).

44. Allen, L. G., Huffman, J. H., Tolman, R. L., Revankar, G. R., Simon, L. H., Robins, R. K., and Sidwell, R. W., 14th Interscience Conference on Antimicrobial Agents Chemotherapy, (1974).

45. Karran, P., and Lindahl, T., Biochemistry 19:6005 (1980).

46. Cunningham, K. G., Hutchinson, S. A., Manson, W., and Spring, F. S., J. Chem. Soc., 2299 (1951).

47. Baker, D. C., Haskell, T. H., and Putt, S. R., J. Med. Chem. 21:1218 (1978).

48. Baker, D. C., Haskell, T. H., Putt, S. R., and Sloan, B. J., J. Med. Chem. 22:273 (1979).

49. Kimchi, A., Shure, H., Lapidot, Y., Rapoport, S., Panet, A., and Revel, M., FEBS Lett. 134:212 (1981).

SYNTHESIS OF PYRROLO[2,3-d]PYRIMIDINE NUCLEOSIDES BY PHASE TRANSFER GLYCOSYLATION AND THEIR FUNCTION IN POLYNUCLEOTIDES

Frank Seela*, Heinz-Dieter Winkeler, Johann Ott, Quynh-Hoa Tran-Thi, Doris Hasselmann, Doris Franzen, and Werner Bußmann

Laboratory of Bioorganic Chemistry, Department of Chemistry, University of Paderborn, D-4790 Paderborn, Federal Republic of Germany

Besides the common nucleosides, more than 50 rare nucleosides have been isolated to date from natural sources. They are found both as monomers in the cultural filtrates of certain microorganisms[1], and also as the rare constituents of nucleic acids[2]. Transfer ribonucleic acids (tRNAs) in particular sometimes contain more than 15% of modified nucleosides[3].

Some years ago we became interested in the synthesis of pyrrolo[2,3-d]pyrimidine nucleosides and their function in polynucleotides.

Queuosine 1

Tubercidin 2   Toyocamycin 3   Sangivamycin 4

Pyrrolo[2,3-d]pyrimidine nucleosides can be divided into two groups. One group covering the antibiotics tubercidin (2), toyo-camycin (3) and sangivamycin (4) is structurally derived from adenosine; these nucleosides have been isolated from natural sources as monomers. Synthetic investigations on these compounds were undertaken by the research teams of R. K. Robins, and L. B. Townsend, and by others. The second group, which is structurally related to guanosine, is found in nucleic acids. Recently queuo-sine (1) was detected in tRNA^{His}, tRNA^{Asp}, and tRNA^{Apn} from E. coli. It is present in the wobble position of these tRNAs and was isolated by S. Nishimura[4]. Its total synthesis was accomplished by T. Goto[5].

Apart from queuosine other 7-deazaguanosine derivatives such as Q*, pre Q_O or pre Q_1[6] are found in tRNAs. These nucleosides are of special interest, since it was shown that a guanine resi-due located in the first position of the anticodon of several tRNAs can be replaced by a 7-deazaguanine moiety. This exchange reaction is catalysed by a transglycosylase[7]. It is suggested that the queuine of mammals is derived from microorganisms such as E. coli[8]. The chemical basis for the transglycosylation reac-tion is the highly stable N-glycosylic bond of pyrrolo[2,3-d]py-rimidine nucleosides, which can shift the equilibrium from a pu-rine towards a pyrrolo[2,3-d]pyrimidine nucleotide.

7-Deazaguanosine   5   Pre Q_1 6   Pre Q_O 7

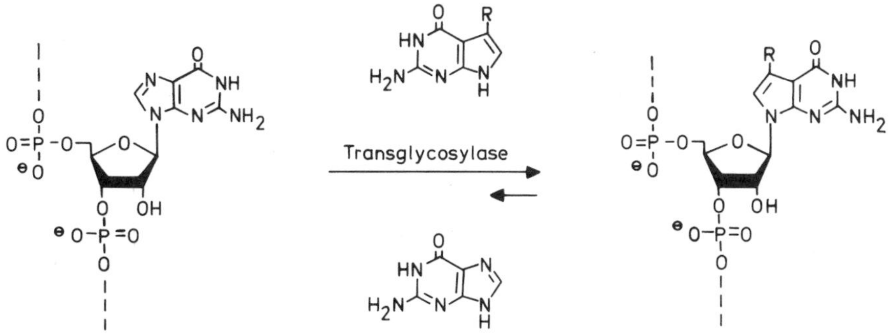

FIGURE 1. Enzymatic transglycosylation of a guanine residue in the first position of the anticodon of tRNA by queuine or the pre Q base (R = CH_2NH_2).

Until now the parent nucleoside of queuosine (1), 7-deazaguasine (5) has not been detected in living systems. Since antimetabolic function of this compound was expected we investigated new routes for its efficient synthesis. Moreover, it is a candidate for mechanistic studies on nucleoside converting enzymes and on the function of pyrrolo[2,3-d]pyrimidine nucleosides in polynucleotides with regard to base recognition and protein synthesis.

The Synthesis of 7-Deazaguanosine by Phase-Transfer-Glycosylation

In 1976 a synthesis of 7-deazaguanosine (5) was reported[9]. Unfortunately the glycosylation under Wittenburg conditions was not regiospecific and it was therefore difficult to prepare the nucleoside in large amounts.

From the course of synthesis of queuosine it was obvious that the generation of a pyrrolo[2,3-d]pyrimidine anion was necessary for a regiospecific 7-glycosylation. As described by Goto et al.[5] the pyrrolyl anion was formed under anhydrous conditions by the action of sodium hydride. When we repeated these experiments we

found the method laborious and obtained only moderate yields.
Therefore, we tried to develop a novel glycosylation method for
pyrrolo[2,3-d]pyrimidines.

Before we approached the glycosylation reaction we investigated the methylation of 7-deazaguanine in order to check the selectivity of electrophilic reactions of 7-deazaguanine (8). Furthermore we also hoped that the spectroscopic characterization of
the isomeric N-methyl-7-deazaguanines might help us to assign the
position of glycosylation.

Employing an excess of dimethyl sulfate in sodium hydroxide
on 7-deazaguanine[10] several reaction products were formed. The
main isomer obtained was the N-3 methyl derivative 10 along with
amounts of the N-1 isomer 11 and the methoxy compound 9. However,
the N-7 methylated isomer could not be detected. These findings
underline the fact that the aglycon 8 is not a candidate for
electrophilic attack under alkaline condition if the formation of
a N-7 substituted reaction product is required.

Although the N-1 methyl isomer 11 was formed only to a small
extent, it is the main product of the condensation reaction of N-
methyl guanidine with ethyl α-cyano-α-(2,3-diethoxy-ethyl)acetate

followed by cyclization of the pyrimidine derivative 14. This reaction is highly stereoselective, probably due to differences in
the nucleophilicity of the methylated and the non-methylated
guanidine nitrogens.

A regioselective 7-methylation was finally achieved when the
protected derivative 16 was chosen. However, it was necessary to
use strong alkaline conditions in order to generate the corresponding pyrrolo[2,3-d]pyrimidine anion. By employing different
methods of methylation we found that methyl iodide in a biphasic
mixture of dichloromethane / 50% aqueous sodium hydroxide in the
presence of a quaternary ammonium salt gave the best results and
led to a regiospecific formation of 17. After hydrolysis of the
methoxy group of 17 the methylthio group of 18 was converted into
an amino group resulting in the formation of the N-7 methyl isomer 19 in high yield.

The isomers 10, 11 and 19 could not be assigned by their UV
spectra. As can be seen from Table 1 (pH 7.0 and 1.0) the N-7 methyl isomer 19 gave spectra with the most bathochromic shift,
which was not expected and is different from other pyrrolo[2,3-d]
-pyrimidines. However, at pH 13.0 the N-1 methyl isomer 11 gave
the expected strongest bathochromic shift. Another difficulty was
the almost identical chromatographic behaviour of the N-3 and the
N-7 methyl isomers. They can, however, be identified by ^{13}C NMR

spectroscopy. As can be seen from the selected data (Table 1) all
three isomers exhibit different NMR data.

The final assignment of all isomers is not discussed here but
will be published elsewhere[11].

TABLE 1.   UV and ^{13}C NMR data of the monomethyl isomers of
7-deazaguanine

		11	10	19
UV:	pH 1	259	262	265
(nm)	pH 7	259	259,(280)	264,(280)
	pH 13	277	259,(280)	267
^{13}C NMR:	N-CH$_3$	33.2	27.5	30.8
(δ)	C-2	150.8	152.6	152.5
	C-6	118.8	116.7	121.2
	C-7a	138.0	149.0	150.0

From the methylation studies it was clear that 7-deazaguanine
(8) could not be used for a regiospecific 7-glycosylation. This
is due to the highly reactive lactam moiety, which enforces a
substitution reaction at the pyrimidine system. However, the protected chromophore 16[10, 12] seemed to be the molecule of choice
for a regiospecific glycosylation under alkaline conditions.
Since we were planning to carry out the glycosylation reaction in
a biphasic mixture of dichloromethane / 50% aqueous sodium hydroxide in the presence of a phase transfer catalyst[13] we selected
the halogenose 20 for glycosylation. The latter was prepared
according to a procedure described by Barker and Fletcher[14] and
was used because of the benzyl protecting groups, which are
stable under alkaline reaction conditions. As a result of the
non-participating 2'-protecting group of the halogenose 20 the
formation of the anomers 21 and 22 was expected.

We were able to glycosylate compound 16 with the halogenose
20 regiospecifically at N-7[15]. Phase transfer glycosylation was
carried out in a reaction vessel depicted in Figure 2.

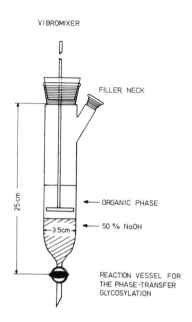

VIBROMIXER

FILLER NECK

25 cm

ORGANIC PHASE

3.5cm

50 % NaOH

REACTION VESSEL FOR
THE PHASE-TRANSFER
GLYCOSYLATION

FIGURE 2.   Reaction Vessel

A rapid transfer of ions through the phase boundary was accomplished by mixing the phases with a vibromixer. This device
allowed immediate formation of the anion of the heterocycle 16
(Figure 3). Apart from the chromophore, the reaction mixture contained tetrabutylammonium bisulfate as the phase transfer catalyst. The tetrabutyl ammonium cation which is soluble in the organic phase replaces the sodium ion during glycosylation and enhances the reaction rate. Glycosylation took place when the halogenose 20 was added (Figure 3); by thorough mixing the reaction
was complete within a few minutes at room temperature. The anomers 21 and 22 shown in the reaction scheme were formed almost
exclusively and in high yield.

The chromatographic separation of the anomers was achieved by
silica gel chromatography with chloroform/methanol as solvent.
The structures of the anomers were assigned by conversion of the
β-anomer 21 into 7-deazainosine (24)[15]. For this purpose the
methoxy group of 21 was cleaved with hydrochloric acid resulting

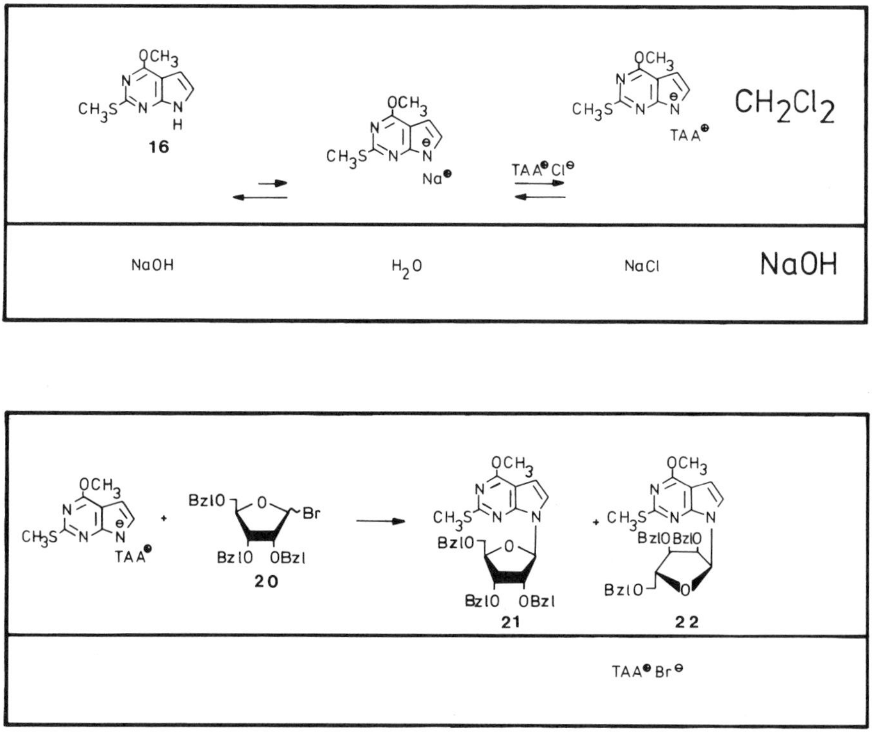

FIGURE 3.   Scheme of phase-transfer glycosylation of the chromophore 16 with the halogenose 20

in the pyrimidinone 15. Desulfurization of 15 using Raney nickel as catalyst gave 23. For the deprotection of the sugar hydroxyls catalytic hydrogenation in the presence of Pd/charcoal was chosen. 7-Deazainosine (24) synthesized by this reaction sequence was identical in all respects with 7-deazainosine obtained by deamination of the antibiotic tubercidin.

During the elucidation of the appropriate reaction conditions for the phase-transfer glycosylation the concentration of NaOH in the inorganic phase was varied between 10% and 50%[16, 17]. In these experiments all other parameters were held constant. As shown in Table 2 a low concentration of sodium hydroxide resulted in a low yield of glycosylation products and a large amount of

unreacted chromophore. This decrease can be explained by solv-
ation phenomena of the ion pair formed between the anion of compound 16 and the tetrabutylammonium counterion. If the NaOH-con-

centration is low, the amount of free water is high. It is then
likely that the anion of the aglycon is encumbered by water molecules and that its reactivity is reduced. In contrast, a relativ-
ly solvation-free anion accompanied by a highly lipophilic sol-
vated tetrabutylammonium cation is generated at a NaOH-concentration of 50%, and therefore its reactivity is high. A highly
reactive aglycon is a basic requirement for an effective glycosylation since the halogenose 20 gradually decomposes under the
strong alkaline reaction conditions of phase-transfer glycosylation.

The ratio of anomers 21 and 22 was also influenced by a
change in the NaOH-concentration (Table 2). Their formation can
be explained by S_N2-displacement of the anomeric bromine substituents of the halogenoses 20 a,b by the anion of the aglycon 16.
Since the β-halogenose 20b formed from either of the anomers of

TABLE 2.

Yield of reaction products (%) formed during phase-transfer glycosylation of 4-methoxy-2-methylthio pyrrolo[2,3-d]pyrimidine with 1-bromo-tri-O-benzyl-D-ribofuranose in dependence of the concentration of sodium hydroxide

| | aq. NaOH | | | | |
	10 %	20 %	30 %	40 %	50 %
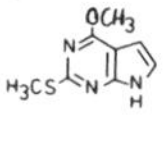 (structure)	65	49	25	17	9
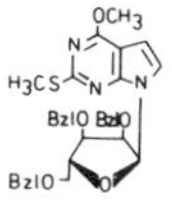 (structure)	10	20	39	55	66
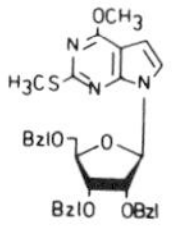 (structure)	25	31	36	28	25
Yield β + α	35	51	75	83	91
Ratio of anomers (β : α)	2.4	1.6	0.9	0.5	0.4

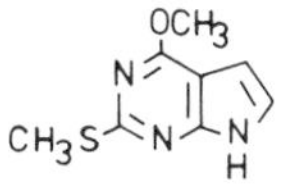

base

16

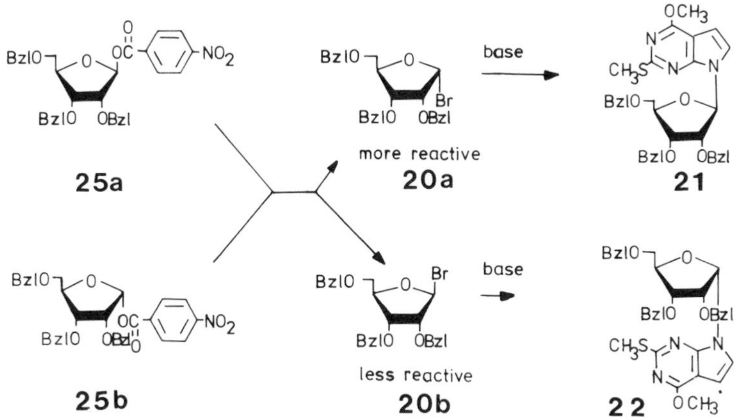

25a

more reactive
20a

21

25b

less reactive
20b

22

p-nitrobenzoyl sugars 25a,b is the dominating anomer, a pref-
erredformation of the α-nucleoside 22 agrees with this mechanism.

The change in the ratio of anomers by decreasing the concentration of NaOH in the inorganic phase is unexpected. As can be
seen from Table 2 the yield of the β-anomer of the nucleoside 21
was almost unaffected by decreasing the NaOH-concentration,
whereas the formation of the α-nucleoside was markedly reduced.

A simple explanation for these phenomena is a preferred decay
of the less reactive β-halogenose 20b by substitution or elimination due to the strongly alkaline reaction conditions. This
side reaction would continously decrease the amount of the β-halogenose 20b resulting in a lower yield of the α-nucleoside 22.
High total yields of phase transfer glycosylation are therefore
only obtained at the highest possible NaOH-concentration in the
inorganic phase. Although aqueous conditions can be used during
glycosylation the amount of "free" water available in the organic
phase should be as low as possible.

In the experiments described above the ratio of catalyst/
chromophore was always about 0.2. If equal amounts were applied

we observed a side reaction of the chromophore itself. This side
reaction was caused by the participation of the solvent CH_2Cl_2
during phase transfer glycosylation. It could be shown that dichloromethane was activated under the strong alkaline reaction
conditions and the halide displaced by the anion of the nucleobase. However, a monosubstitution product was not detected. It is
therefore assumed that in a fast reaction HCl was eliminated
forming a methylene imonium ion as an intermediate. This cation
immediately attacks another chromophore anion and connects two
chromophors via a methylene bridge. By omitting the halogenose 20
we isolated the bridged nucleobases 27 in more than 50% yield,
when high amounts of catalyst were chosen and the reaction time
was extended to 2 h[18]. The novel 7,7'-methylenedipyrrolo[2,3-d]-
pyrimidines 26 and 27 are interesting model compounds for the
studies of stacked and nonstacked nucleobases.

The conversion of the 2-methylthio group of 21 into an amino
group was carried out under conditions which have been already
used for the synthesis of queuosine[5]. The 4-methoxy group of 21
was first cleaved by acidic hydrolysis as already described. This
afforded the pyrimidinone 15. Nucleophilic displacement could
then be performed selectively at C-2 without affecting C-4. In
order to avoid anion formation at N-3 or O-4 under the alkaline

conditions necessary for this displacement reaction, the nitrogen-
3 was protected with an isopropoxymethyl residue. Compound <u>15</u> was
alkylated under anhydrous conditions with isopropylchloromethyl
ether in the presence of sodium hydride to give <u>28</u>. After chromatographic purification the material was treated with a molten
mixture of acetamide and sodium hydride yielding the protected 7-
deazaguanosine derivative <u>29</u>. Because of the stability of the N-
glycosylic bond in pyrrolo[2,3-d]pyrimidine nucleosides the
cleavage of the N-3 isopropoxymethyl group of <u>29</u> was attempted by
hydrochloric acid. However, all efforts to obtain the N-3 deprotected compound failed. After evaporation of the solvent at least
four reaction products were   detected. None of these products
showed the typical UV spectrum of the aglycon with an absorption
maximum at 260 nm and a shoulder around 280 nm. Because of these
difficulties we then gave preference to the removal of the benzyl
protecting groups with boron trichloride.

Treatment of the protected nucleoside <u>29</u> with boron trichloride in dichloromethane at -78°C removed the benzyl protecting
groups. After work up of the reaction mixture we obtained a material which did not show the UV spectrum of a 3,7-protected agly-

con. Instead an UV spectrum was observed, which was similar to 2-acetamino-3,7-dihydro-pyrrolo-4H-[2,3-d]pyrimidine-4-one. Therefore the application of boron trichloride removed not only the benzyl protecting groups but also the isopropoxymethyl residue. After preparing compound 30 in larger amounts we removed the acetyl group by treatment with aqueous ammonia/methanol and obtained 7-deazaguanosine (31) which crystallized from methanol or water in colorless needles[19]. The α-anomer 32[16] was prepared by the same procedure as described for the β-anomer 31.

The UV spectrum of the nucleoside was very close to that of 7-deazaguanine and did not show a strong dependence of the UV-maximum on the pH change. On TLC it showed similar behaviour to guanosine. The final proof of the structure of 7-deazaguanosine (31) and its α-anomer 32 was made on the basis of ^{1}H and ^{13}C NMR spectroscopy.

The Mononucleotides of 7-Deazaguanosine

The synthetic sequence for preparation of mono- and polynucleotides of 7-deazaguanosine (c^7G, 31) started with regioselective phosphorylation. Phosphorylation of compound 31 with $POCl_3$
in trimethyl phosphate yielded the monophosphate 33 which was purified by anion exchange chromatography on DEAE cellulose. The
conversion of the monophosphate into the diphosphate 35 was accomplished according to the method of Hoard and Ott[20]. Activation
of the tri-n-butylammonium salt of c^7GMP with 1,1'-carbonyldi-
imidazole yielded the imidazolidate 34 and condensation with tri-
n-butylammonium phosphate gave compound 35. Its structure was
confirmed by its UV spectrum coinciding with that of c^7G and doublets at -4.99 and -9.88 ppm with a coupling constant of 22 Hz in
the ^{31}P NMR spectrum.

Nucleoside cyclophosphates play an important role in the metabolism of the cell. In particular the 3',5'-cyclophosphate of
guanosine (cGMP) occupies a key position in the regulation of
hormone activity; furthermore it activates a series of enzymes,
e.g. protein kinases. Recently it has been established that cGMP

is also of great importance in vision: photoactivation of one
rhodopsin molecule leads to enzymatic hydrolysis of 10^5 cGMP molecules. The negative charge of the cyclonucleotides hinders their
penetration of the cell membrane. Thus, incubation of intact
cells with these compounds increases the activity only to a small
extent. In consequence cyclonucleotides with more lipophilic
groups are of interest. Pyrrolo[2,3-d]pyrimidine nucleosides are
more lipophilic than the corresponding purine compounds, since
they do not contain a nitrogen in the purine 7-position. Furthermore
the pyrrolo[2,3-d]pyrimidine nucleosides show a decreased tendency
to aggregate, thus incrasing their availabitity in aqueous solution. This prompted us to synthesize the cyclophosphate 36.

The monophosphate 33 was cyclized with N,N-dicyclohexylcarbo-
diimide in pyridine yielding the cyclophosphate 36 in high yield.
Subsequent anion-exchange chromatography and lyophilization produced pure 36 as the amorphous triethylammonium salt. The structure of compound 36 was confirmed by its ^{1}H NMR spectrum showing
a singlet for the anomeric proton[21].

33   **36**

Cyclophosphate-specific phosphodiesterase cleaved the cyclonucleotide 36 to the 5'-monophosphate 33. Table 3 shows the K_M
and V_{max} values of the 7-deazanucleotide 36 compared to those of
cGMP.

The 7-purine nitrogen is thus not a requirement for the binding of cGMP and its analogues to the phosphodiesterase and for
their hydrolysis by this enzyme, as is indicated by the similarity of the K_M values of cGMP and its 7-deaza analogue. The higher
V_{max} value of cyclo c^{7}GMP compared to cGMP may be due to an in-

creased ring strain in the 7-deaza cyclophosphate residue induced
by the chromophore through the highly stable N-glycosylic bond.
Furthermore at concentrations above 1 mM an apparent substrate
inhibition of cyclo c^7GMP was observed which has also been found
for cGMP.

TABLE 3. K_M and V_{max} values of cyclonucleotide phosphodiesterase for cyclo c^7GMP and cyclo GMP. The reaction mixture
(1 ml) contained phosphodiesterase from bovine heart (12.2 μg),
50 μmol Tris-HCl, pH 7.5, 10 μmol $MgCl_2$ and the corresponding
cyclo nucleotide.

compound	K_M [mM]	V_{max} [mM/min·mg protein]
cyclo c^7GMP	0.24	7.3
cyclo GMP	0.15	1.4

The nucleoside pyrophosphate 35 was the starting material for
the synthesis of poly(7-deazaguanylic acid) by polynucleotide
phosphorylase.

Poly(7-deazaguanylic acid)

Although polynucleotide phosphorylase from M. luteus catalyzes the polymerisation of a number of ribonucleoside diphosphates,[22] the synthesis of high molecular weight poly(G) only proceeds with difficulty under the usual conditions. The synthesis
of this polymer can, however, be achieved by using polynucleotide
phosphorylase from E. coli at elevated temperatures (60°C) and in
the presence of Mn^{2+} instead of Mg^{2+}. In contrast, the polymerisation of c^7GDP proceeds instantaneously even with polynucleo-

tide phosphorylase from M. luteus in the absence of manganese
ions under regular conditions giving poly(c^{7}G) in about 50% yield.

An explanation for this difference may be derived from the
observation that c^{7}G and its phosphates show no tendency to form
gels in aqueous solution as is found for GMP[24]. This implies
that the lack of N-7 prevents the self-aggregation which   characteristic of guanosine phosphates. Consequently there will be a
more ready availability of the substrate at the active center of
polynucleotide phosphorylase. Furthermore the polynucleotide
poly(c^{7}G) may not form such strong aggregates as does poly(G). It
should be also noted that tubercidin was even more readily incorporated into polymers by M. luteus PNPase than was ADP[25].

The UV spectrum of poly(c^{7}G) at pH 7.0 shows a maximum at
258 nm with a shoulder around 278 nm (Figure 4). The shape of the
absorption band is similar to that of poly(G), but the maximum
shows a bathochromic shift of 5 nm (Table 4). The absorbance in
0.1 N HCl is decreased, whereas in 0.1 N NaOH it shows an increase. These changes are not observed for c^{7}GMP. The increase in
absorbance under alkaline conditions can be interpreted as an un-
stacking of the nucleobases by anion formation. This is under-
lined by pK measurements. From the difference in the absorbance

at 275 nm between the neutral and the anionic forms of c[7]GMP the pK value of base deprotonation was determined to be 10.7.

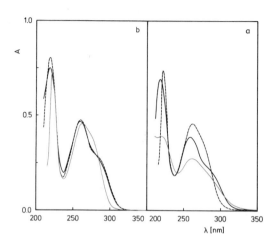

FIGURE 4. UV spectra of poly(c[7]G) (a) and c[7]G (b) in 0.07 M phosphate buffer, pH 7.0 (——), 0.1 M HCl (···), and 0.1 M NaOH (---) at 25°C.

TABLE 4. UV Data and Hypochromicities[α] of Poly(c[7]G) and Poly(G)

	poly(c[7]G)	poly(G)
UV_{max} (nm)		
pH 7.0	258, 278 (sh)	253
pH 1.0	260	256, 278
pH 13.0	261	260, 268
ε_{max} (pH 7.0	10 300	10 100
hypochromicity by phosphate determination (%)	22	26
hypochromicity by nuclease T_1 treatment (%)	21	
thermal hypochromicity between 20° and 90°C (%)	11	12

[α] Hypochromicities are measured at the long wavelength maximum.

The hypochromicities of poly(c^{7}G) and poly(G) were determined
by the molybdenum blue method[26]. When extinction coefficients of
13,200 for c^{7}G and 13,600 for G were used, hypochromicities of
22% and 26% were found (Table 4). The hypochromicity of poly(c^{7}G)
was therefore very close to that of poly(tubercidylic acid),
which exhibited 23%[25].

Poly(c^{7}G) is similar to poly(G) in that it shows no hypochro-
mic change during heating under high-salt conditions and in the
presence of magnesium ions, indicating a rigid structure. If di-
valent ions were removed by dialysis against EDTA and the change
of absorbance was measured in 0.07 M phosphate - 1 mM EDTA at
pH 7.0, thermal melting was found. Figure 5 shows the melting
profile of poly(c^{7}G) and poly(G) under identical salt concentrations and in the absence of magnesium. Whereas poly(G) showed a

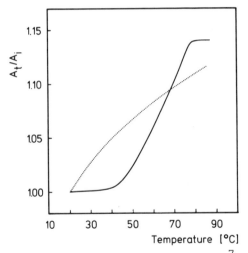

FIGURE 5.   Denaturation profiles of poly(c^{7}G)($\cdots$) and
poly(G)(——) in 0.07 M phosphate buffer-1 mM EDTA (pH 7.0). A_t/A_i
is the ratio of the λ_{max} absorbances at a given temperature (t)
to that at the initial temperature (i).

sigmoid denaturation curve indicating strong cooperativity of
this process, the curve of poly(c^{7}G) showed no inflection point.

The same has been found for poly(c[7]I)[27]. While poly(G) forms a tetrastranded helix[28] by Hoogsteen base pair formation, involving N-7, this is not possible for poly(c[7]G) because of the lack of N-7. This implies that poly(c[7]G) will only exist as a single-stranded polymer under these conditions with an increased flexibility of the polynucleotide chain. A more flexible structure of poly(c[7]G) compared to that of poly(G) is also indicated by the CD spectra[23].

The stoichiometry of the complex poly(c[7]G)·poly(C) (Figure 6) was determined by the method of continuous variation[29]. A well-defined absorbance minimum was obtained at 290 nm. This minimum was reached at exactly 50 mol % poly(c[7]G), clearly demonstrating

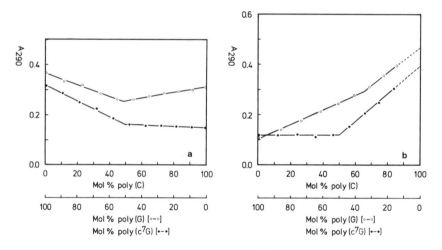

FIGURE 6. Mixing curves of poly(c[7]G) and poly(G) with poly(C) at 290 nm in 0.1 M Tris-HCl buffer, pH 7.0, 0.2 M NaCl, and 1 mM MgCl$_2$ (a) and the same compounds at 290 nm in 0.1 M glycine-HCl buffer, pH 2.5, containing 0.2 M NaCl and 1 mM MgCl$_2$ (b).

a 1:1 complex at pH 7.0. The same stoichiometry is observed for the complex of poly(G) and poly (C) under identical conditions. A definite break in the mixing curves of poly(G)·poly(C) was only observed if the mixture were allowed to come to equilibrium for 2 h. In contrast to poly(G), an immediate complex formation of poly(c^{7}G) with poly(C) was observed. The decreased rate of poly(G)·poly(C) formation can be explained by the necessary dis-aggregation ot the tetrastranded poly(G); this does not occur in the case of poly(c^{7}G) because of its single-stranded structure.

The protonation of N-7 of poly(G) at acidic pH is a require-ment for Hoogsteen base pair formation with poly(C). In agreement with this formation a 1:2 stoichiometry was observed at pH 2.5 of poly(G)·2poly(C) (Figure 6).

In contrast to poly(G), triplex formation of poly(c^{7}G) with poly(C) did not occur even under acidic conditions. As Figure 6 shows, the minimum of the mixing profile was still located at 5O mol % poly(c^{7}G) showing that poly(c^{7}G) can only form Watson-Crick base pairs and the triple-strand formation of poly(G) defi-nitely occurs via N-7.

At pH 5.3 in EDTA the complex of poly(G)·poly(C) melts coop-eratively (Figure 7) with a T_m of 75°C. Surprisingly the complex of poly(c^{7}G)·poly(G) shows an almost identical melting point of T_m = 74°C. These data imply that the stereochemical arrangement of c^{7}G in a base paired Watson-Crick double helix is very similar to that of G, a fact that would allow us to consider why 7-deaza-nucleotides are utilized to modulate polynucleotide recognition.

The structure of Watson-Crick base paired polynucleotides can also be detected with the enzyme nuclease S_1[30]. This enzyme is single-strand specific and recognizes secondary structures of nucleic acids. Double-stranded 2'-deoxyribo- and ribopolynucleo-tides are not significantly cleaved by the enzyme – the same is true for other polynucleotides forming multistranded structures. Consequently poly(G) is not cleaved by nuclease S_1 (Figure 8). However, as the figure shows this enzyme hydrolyzes poly(c^{7}G) immediately. These findings are in agreement with the results re-ported above that poly(c^{7}G) does not form multistranded struc-

tures, because of its inability to bind another strand via Hoogsteen base pairs resulting from the lack of N-7.

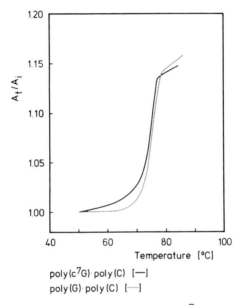

poly(c^{7}G)·poly(C) [—]
poly(G)·poly(C) [·····]

FIGURE 7. Melting curves of poly(c^{7}G)·poly(C) (——) at 265 nm
and poly(G)·poly(C) (···) at 257 nm in 0.2 mM EDTA (pH 5.3). A_t/A_i
is the ratio of absorbance at a given temperature (t) to that at
the initial temperature (i).

The action of ribonuclease T_1 on polynucleotides has been the
subject of intensive study. The enzyme hydrolyzes the internucleotide bond of RNAs adjacent to the 3'-phosphate of GMP. It has
been reported that N-7 of guanosine is involved in the binding of
the substrate to the active site of the enzyme[31]. On the basis of
enzyme action on various substrates, a model of the enzyme substrate complex has been proposed. Results on the protonation of
N-7 of guanosine during enzymatic binding were controversial[32].

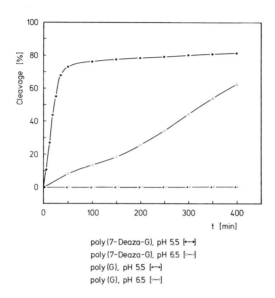

poly (7-Deaza-G), pH 5.5 [•—•]
poly (7-Deaza-G), pH 6.5 [◦—◦]
poly (G), pH 5.5 [▲—▲]
poly (G), pH 6.5 [△—△]

FIGURE 8. Cleavage of polynucleotides as indicated with
nuclease S_1 at pH 5.5 and 6.5, at 25°C. The reaction mixture contained 0.1 μmol/ml of the polynucleotides, 2000 units of the
enzyme, 0.29 M NaCl, 1 m M $ZnSO_4$, and 0.03 M sodium acetate (pH
5.5 or 6.5).

Poly(c^7G) is a ideal probe, to test the proposed model, which is
based on a hydrogen bridge between N-7 of guanosine and the
His-92 residue of the enzyme. We therefore applied RNAse T_1 to
poly(c^7G) and measured the cleavage of the polynucleotide by
observing the change of hypochromicity. At this value (Table 4)
rapid hydrolysis occured, resulting in 80% cleavage within 3 h.
This clearly demonstrates that N-7 of the guanosine residue is
not a requirement for RNAse T_1 action on polynucleotides.

Acknowledgement

We gratefully acknowledge the critical reading of the manuscript by Dr. V. Armstrong. The work was supported by a grant of
the Deutsche Forschungsgmeinschaft.

References

1. Suhadolnik, R. J.,"Nucleoside Antibiotics", S. Wiley-Inter-
   science, New York 1971.

2. McCloskey, J. A. and Nishimura, S., Acc. Chem. Res. 1977,
   10, 403.

3. Sprinzl, M. and Gauss, D. H., Nucleic Acids Res. 1982, 10,
   r1.

4. Kasai, H., Ohashi, Z., Harada, F., Nishimura, S., Oppenhei-
   mer, N. J., Crain, P. F., Liehr, J. G., von Minden, D. L.,
   and McCloskey, J. A., Biochemistry 1975, 14, 4198.

5. Ohgi, T., Kondo, T., and Goto, T., J. Amer. Chem. Soc. 1979,
   101, 3629.

6. Okada, N., Noguchi, S., Nishimura, S., Ohgi, T., Goto, T.,
   Crain, P. F., and McCloskey, J. A., Nucleic Acids Res. 1978,
   5, 2289; Noguchi, S., Yamaizumi, Z., Ohgi, T., Goto, T.,
   Nishimura, Y., Hirota, V., and Nishimura, S., Nucleic Acids
   Res. 1978, 5, 4215.

7. Katze, J. R., and Farkas, W. R., Proc. Natl. Acad. Sci. USA
   1979, 76, 3271.

8. Reyniers, J. P., Pleasants, J. R., Wostmann, B. S., Katze,
   J. R., and Farkas, W. R., J. Biol. Chem. 1981, 256, 11591.

9. Townsend, L. B., Tolman, R. L., Robins, R. K., and Milne, G. H., J.Heterocycl. Chem. 1976, 13, 1363.

10. Davoll, J., J. Chem. Soc. 1960, 131.

11. Seela, F., Götze, A., and Bussmann, W. (in preparation).

12. Seela, F., and Richter, R., Chem. Ber. 1978, 111, 2925.

13. Weber, W. P., and Gokel, G. W., "Phase Transfer Catalysis in Organic Synthesis", Springer-Verlag, West-Berlin 1977; Starks, C. M., and Liotta, C., "Phase Transfer Catalysis", Academic Press, New York 1978; Dehmlov, E. V., and Dehmlov, S. S., "Phase Transfer Catalysis", Verlag Chemie, Weinheim 1980.

14. Barker, R., and Fletcher jr., H. G., J. Org. Chem. 1961, 26, 4605.

15. Seela, F., and Hasselmann, D., Chem. Ber. 1980, 113, 3389.

16. Seela, F., Hasselmann, D., and Winkeler, H.-D., Liebigs Ann. Chem. 1982, 499.

17. Seela, F., and Winkeler, H.-D., J. Org. Chem. 1982, 47, 226.

18. Seela, F., and Menkhoff, S., Liebigs Ann. Chem. 1982, 1405.

19. Seela, F., and Hasselmann, D., Chem. Ber. 1981, 114, 3395.

20. Hoard, D. E., and Ott, D. G., J. Am. Chem. Soc. 1965, 87, 1785.

21. Tran-Thi, Q.-H., Franzen, D., and Seela, F., Angew. Chem. Int. Ed. Engl. 1982, 21, 367.

22. Godefroy-Colburn, T., and Grunberg-Manago, M.,"Enzymes", 3rd Ed. 1972, 7, 533.

23. Seela, F., Tran-Thi, Q.-H., and Franzen, D., Biochemistry 1982, 21, 4338.

24. Fisk, C. L., Becker, E. D., Miles, H. T., and Pinnavaia, T. I., J. Am. Chem. Soc. 1982, 104, 3307.

25. Seela, F., Tran-Thi, Q.-H., Mentzel, H., and Erdmann, V. A., Biochemistry 1981, 20, 2559.

26. Chen, P. S., Toribara, T. Y., and Warner, H., Anal. Chem. 1956, 28, 1756.

27. Torrence, P. F., DeClercq, E., Waters, J. A., and Witkop,
    B., Biochemistry 1974, 13, 4400.

28. Thiele, D., and Guschlbauer, W., Biopolymers 1971, 10, 143.

29. Felsenfeld, G., Biochim. Biophys. Acta 1958, 29, 133.

30. Ando, T., Biochim. Biophys. Acta 1966, 114, 158.

31. Egami, F., Oshima, T., and Uchida, T., Mol. Biol. Biochem.
    Biophys. 1980, 32, 250.

32. Oshima, T., and Imahori, K., in "Proteins - Structure and
    Function" 1971, Vol. I, Kodansha, Tokyo, pp. 333.

THE ALKYLSILYL PROTECTING GROUPS:IN PARTICULAR,
THE T-BUTYLDIMETHYLSILYL GROUP IN NUCLEOSIDE AND
NUCLEOTIDE CHEMISTRY

Kelvin K. Ogilvie

Department of Chemistry
McGill University
Montreal, Quebec, Canada

ABSTRACT

The alkylsilyl protecting groups have made a remarkable
impact on the practicality of oligoribonucleotide synthesis.
Some of the reagents developed are stable to acid and base but
are removed under nearly neutral conditions. They have provided
synthetically useful derivatives which can be analyzed directly
by gas-liquid chromatography and mass spectrometry. While the
triisopropylsilyl and t-butyltetramethylenesilyl are the most
stable, the commercial availabiliy of the t-butyldimethylsilyl
group has made it the reagent of choice. All silyl groups are
rapidly cleaved by fluoride ion. We have in turn thoroughly
investigated all possible degrees of silylation of deoxynucleosides, ribonucleosides and most recently, arabinonucleosides.

209

The silylating reagents do not react with amino groups on nucleosides nor do the chlorophosphite reagents used in formation of
the internucleotide linkage. This has led to a remarkably efficient overall synthesis of ribonucleotides and arabinonucleotides. The combination of methoxytrityl or dimethoxytrityl,
t-butyldimethylsilyl, and levulinyl groups gives a complete
protection system.

INTRODUCTION

Nucleic acids contain three important sections: the heterocyclic bases, the sequence of which determines the information
carried by the nucleic acid; the sugar or carbohydrate portion
and the phosphate linkers which join the carbohydrate units
together. The combination of a carbohydrate unit and a heterocyclic base is called a nucleoside. A nucleoside having a phos-

phate group on any hydroxyl of the carbohydrate unit is called a
nucleotide.   Important nucleic acids, DNA and RNA, are therefore
polynucleotides. The major difference between DNA and RNA occurs
at the carbohydrate portion of the nucleoside units. The deoxynucleosides which make up DNA ($\underline{1}$, X=H) are made from 2-deoxy-
ribose, while RNA ($\underline{2}$, S=OH) is made from nucleosides possessing
the ribose structure. The only other significant composition
difference between DNA and RNA occurs at the bases. Both DNA and
RNA use the purines adenine (Ad) and guanine (Gu) and the pyrimidine cytosine (Cy). DNA incorporates the pyrimidine thymine (Th)
while RNA incorporates uracil (Ur). Until recently, it had been
assumed that all natural nucleic acids were composed of 3' to 5'-
phosphate bridges and this is true for the molecules tradition-
ally referred to as DNA and RNA. However, it has been discovered[1-3] that short oligoadenylates possessing 2' to 5' linkages play a key role in the interferon process.

adenine   guanine

cytosine   thymine   uracil

The objective in the chemical synthesis of oligo- and polynucleotides is simply to link nucleoside units together via the

correct phosphate bridge. To date all successful chemical approaches have started with a protected nucleoside or mononucleotide as the key starting material.

In order to successfully link nucleosides together two critical conditions must be met: (1) protecting groups must be available for nucleoside units and (2) efficient phosphorylative condensation procedures must exist. These two conditions have been the major focus of research in this area for over two decades.

Most current strategies for oligonucleotide synthesis involve the triester method. The first successful and practical triester synthesis was introduced by Letsinger[4] in the 60's. The objective of the triester procedure is to provide, at the end of each condensation sequence, a nucleotide unit that is fully protected at least at the carbohydrate and phosphate moieties. A typical product[5a] of condensing two ribonucleosides (eg. 3 and 4) is unit 5. An essential requirement of this unit is that it must be capable of undergoing extension from either the 3'-or 5'-end. This means that R and R''' must be capable of being removed independently of one another and of R' and R''. If additional protecting groups are required for amino groups on the base rings B, the problem can take on added complexity.

A key aspect of the problem of ribonucleoside protection is the nature of R' which must be chosen with great care. If R' is base labile (eg. acyl) it must be sufficiently stable such that migration between the 2'- and 3'-positions is slower than the rate of phosphorylation at the 3'-position. However, R' must not

be so stable that the basic conditions required to remove it will cause cleavage of the resulting ribonucleotide. On the other

hand if R' is acid sensitive, it must not require acid conditions
that result in isomerization of the phosphate linkage from the
3'- to the 2'-position. These problems with acyl and acid labile
(eg. tetrahydropyranyl) groups have been dealt with in detail by
Reese.[5b]

Let us assume that a group R' has been found which satisfies
the above requirements and is stable to the conditions that lead
to the removal of R and R'''. The problem is now reduced to
introducing R' on the 2'-position of the nucleoside. (We will
use for illustration the problem of synthesizing a 3'-5'-linked
ribonucleotide. The problems in the synthesis of 2'-5'-linked
nucleotides are obviously similar and complementary and will be
dealt with later for specific examples.)

For the moment we will assume that units of the type 8 are
readily available. The introduction of R' simply requires a

reaction between R'X and the 2'-position of 8. Since the 2'- and 3'-positions are cis, secondary hydroxyls with little difference in pKa's, nearly equal amounts of 3 and 9 would be expected. Perhaps for the situation where R is very bulky and including the fact that the 2'-OH is generally slightly more acidic than the 3'-OH, one might predict a slight excess of 3 to be produced. This is generally the case. What is also true, is that 3 and 9 usually have very similar chromatographic properties making practical separation of large quantities very difficult.

DISCUSSION

It is clear from the above discussion that protecting ribo-nucleosides for ribonucleotide synthesis is not an easy problem. The choice of R' is the key to the solution. This latter problem has been on our minds for years We[6] (and others[7,8]) initially thought that the best way to deal with the 2'-OH was to eliminate it temporarily and reintroduce it later. Anhydronucleosides (or cyclonucleosides) of the type 10 seemed ideal. Such compounds resemble deoxynucleosides for which protection of the 3'- and 5'-positions is relatively easy. Furthermore, the ribo structure could be reintroduced by treatment with sodium benzoate. While this route proved useful[6] in the synthesis of analogues, includ-ing arabinonucleotides, of natural nucleic acids it has not yet been shown to be practical for the synthesis of ribonucleotides.

$$\underset{10}{\text{10}}$$

Our attention was then directed toward more conventional protection methods and our objectives were raised slightly by the addition of another goal -- to obtain a protecting group which would be essentially stable to acid and base and that could be removed under neutral conditions. The solution described in this report was arrived at when we considered another objective - to render synthetically useful protected nucleosides volatile enough for gas chromatography and mass spectrometry.

The normal route to this latter end is to trimethylsilylate[9] polar materials which are then sufficiently volatile for analysis. Of course the trimethylsilyl group is not sufficiently stable to most of the conditions employed during nucleotide synthesis. However, in carbon compounds increasing the steric bulk at the reaction center usually reduces the rates of solvolysis. This trend had been shown to exist in silyl ethers by the work of Gilman[10] and others[11] in the late 40's and early 50's. These workers had prepared a number of bulky silyl chlorides. Stork[12] had used one of these, the t-butyldimethylsilyl (TBDMS) group in the protection of enols in 1968. Furthermore, Hancock[13] had used tri-n-butylchlorosilane and triphenylchlorosilane for derivatizing nucleosides for gas chromatography. We chose several different alkylsilyl groups for investigation and the initial results, along with the timely appearance of an article by

Corey[14], convinced us that we were moving in the right direction. The remainder of this review will be used to summarize our development of the alkylsilyl protecting groups, particularly the t-butyldimethylsilyl (TBDMS) group in nucleoside and nucleotide chemistry.

Our initial reports[15,16] involved the TBDMS group and deoxy nucleosides. The ease of preparation[11] of t-butyldimethylsilyl chloride (TBDMS-Cl) and the fact that deoxynucleosides constitute a more simple model system than ribonucleosides were the basis for this choice. We found[15,16] that TMDMS-Cl was quite selective for the primary 5'-hydroxy giving yields of 12 in the order of 75%. These initial yields are lower than the 90% yields of 12 we reported subsequently.[17] The difference in these yields is

easily accounted for. In our first reports reactions were carried out by placing the solid reagents (eg. nucleoside, TBDMS-Cl and imidazole) in a flask followed by addition of solvent (DMF). While all the TBDMS-Cl and imidazole dissolved instantly, the nucleoside usually required 1-4 min to completely dissolve. Consequently, the initial ratio of TBDMS-Cl:11 was much greater than 1.1:1.0. Due to the high reactivity of TBDMS-Cl, much more 14 was produced than expected. By adding silylating agent to

completely dissolved nucleoside the yields of 12 rose to ~90% in
all cases[17] while the yields of 14 fell to 0-5%.

Silyation of nucleosides in DMF containing added base are
very rapid usually being complete within 15-30 min. Product
composition does not change if the reaction mixture is allowed to
stand for 24 h. Reactions are slower in pure pyridine, often
requiring several hours for completion. The disilyl derivatives
14 are obtained in >90% yields by using an excess of silylating
agent.[16]

The reaction of TBDMS-Cl with all the common deoxynucleosides and the N-acyl derivatives of dA, dC and dG has been described in detail along with the use of these derivatives in
further derivatizing the deoxynucleosides.[16] In the case of the
nucleosides having exocyclic amino groups (dA, dC, dG), derivatives bearing a silyl group on the amino group were not detected.
Even when using an excess of silylating agent such derivatives
could occasionally be isolated in a yield of only a few percent.
It is possible that N-silylated derivatives are formed which are
either unstable to work-up conditions (and TLC) or to the reaction conditions. The amount of silylating agent required to give
good yields of 12 varied with the nucleoside (from 1.1 eq for dT
to 4.4 eq for dG). We have also noted that the activity of the
reagent decreases with age and the number of exposures to air
(moisture). Thus, the exact amount of TBDMS-Cl required in any
experiment depends directly on the quality of the reagent and
thus reactions should be followed by TLC.

While the direct silylation procedure leads to rapid formation of 12 or 14 in excellent yields, the 3'-TBDMS derivatives 13
are most readily available from the methoxytritylnucleosides 15.

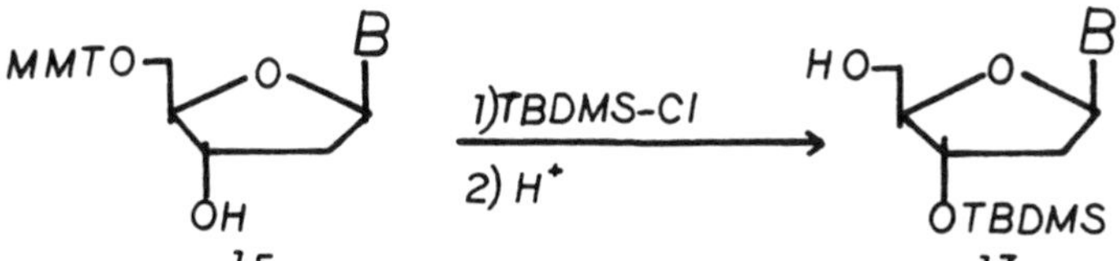

$$\underset{\underline{15}}{MMTO-\boxed{}-B} \xrightarrow[\text{2) } H^+]{\text{1)} TBDMS\text{-}Cl} \underset{\underline{13}}{HO-\boxed{}-B}$$

70−85%

The TBDMS group is reasonably stable to the basic conditions required to remove common 0-acyl groups.[16] The TBDMS group can be removed with 80% HOAc[16,17] and faster from the 5'-position (5-6 h) than from the 3'-position (13, 25 h). This means that the di-TBDMS compounds 14 which are obtained in near quantitative yields can be converted to the 3'-TBDMS derivative (13) in good yields on acid hydrolysis.[16−18]

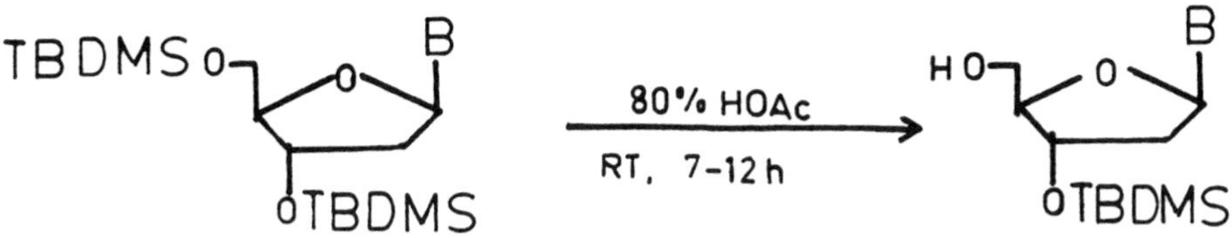

$$\underset{\underline{14}}{TBDMSO-\boxed{}-B} \xrightarrow[\text{RT, 7-12 h}]{80\% \text{ HOAc}} \underset{\underline{13} \quad 60\%}{HO-\boxed{}-B}$$

In order to select silyl groups offering a range of stabilities toward acid hydrolysis we investigated[17] a number of silyl chlorides shown below. All of these reagents (16-20) reacted cleanly with nucleosides. As the steric bulk at silicon increased so did the stability of the resulting nucleoside ethers. The order of stability was TMTBS > TIPS > TBDMS ~ TBMODS > TMIPS > MDIPS. In all of these cases the 3'-isomer was more stable to acid than the 5'-derivative. The t-butyltetramethylene derivatives (from 20) required 0.01N HCl at ~90°C for efficient removal.

We also investigated[17] the use of acetamide and imidazolide derivatives of alkylsilanes as silylating agents. These derivatives gave very clean reactions with nucleosides and are ideal for acid-sensitive nucleosides or their derivatives.

All of the silyl groups described above gave nucleoside derivatives which could be characterized directly by gas chromatography.[17,18] In general, 3'-derivatives have shorter retention times than the 5'-isomer and 3',5'-disilyl derivatives have the longest retention times.

Perhaps the biggest surprise in the properties of silylated nucleosides came from their mass spectra.[19-21] The 5'-silyl derivatives (12) have three important high mass fragments. These occur at $(M-alkyl)^+$, $(M-alkyl-H_2O)^+$ and $(M-alkyl-2H_2O)^+$. The 3'-isomer (13) shows the $(M-alkyl)^+$ ion but the $(M-alkyl-H_2O)^+$ is a very minor peak and the $(M-alkyl-2H_2O)^+$ ion is absent. In the cases of silyl groups where one of the alkyl groups is t-butyl (on silicon), the $(M-alkyl)^+$ fragment is always $(M-\underline{t}-butyl)^+$, ie $(M-57)^+$. The loss of the water molecules from the 5'-isomer occurs sequentially as confirmed by metastable peaks. The difference in fragmentation pattern between the 3'- and 5'-isomers means that structures can be assigned to the derivatives of rare or expensive nucleosides on the basis of very small amounts of material.

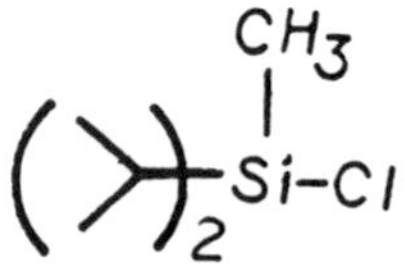

MDIPS-Cl
16

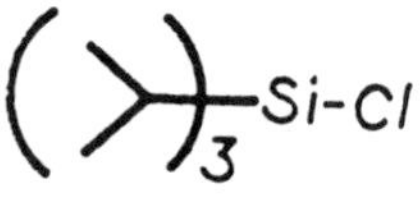

TIPS-Cl
17

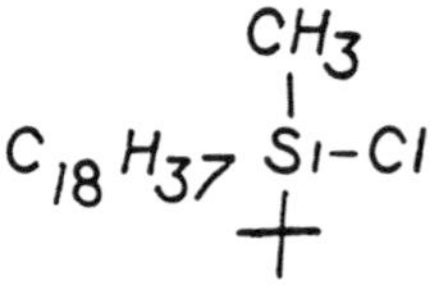

TBMODS-Cl
18

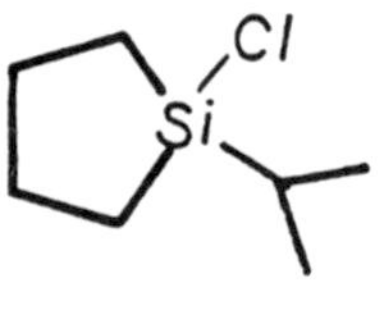

TMIPS-Cl
19

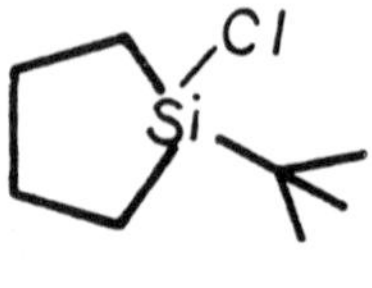

TMTBS-Cl
20

Partial Mass Spectra of Mono-silyl Derivatives of Thymidine (__12__ and __13__)

Silyl Group	(M-alkyl)$^+$			(M-alkyl-H$_2$O)$^+$			(M-alkyl-2H$_2$O)$^+$		
	m/e	*RI$^{5'}$	RI$^{3'}$	m/e	*RI$^{5'}$	RI$^{3'}$	m/e	*RI$^{5'}$	RI$^{3'}$
__a__, MDIPS[b]	327	5.7	9.3	309	36	0.72	291	13	——
__b__, TMIPS[b]	325	0.64	6.9	307	5.5	4.0	289	1.8	——
__c__, TIPS[b]	355	4.2	2.8	337	30	0.13	319	13	——
__d__, TMTBS[b]	325	5.8	15	307	31	—	289	7.8	——
__e__, TBDMS[a]	299	7.0	10.2	281	30	0.5	263	6.2	——

*RI=relative intensitity (to base peaks).

[a] Recorded on Hitachi RMU-6D single-focussing mass spectrometer at 50 eV.

[b] Recorded on Finnigan 1015 quadrupole mass spectrometer at 70 eV and corrected for mass discrimination by assuming ion transmission inversely proportional to m/e. Alkyl is __iPr__ or __tBu__.

The final question to be answered concerning the utility of the silylated nucleosides was their stability toward phosphorylation conditions. We found that the TBDMS group was stable to all of the normal procedures used in the synthesis of oligodeoxynucleotides and several deoxynucleotides were synthesized.[17]

The alkylsilyl groups can all be removed by fluoride ion. The reagent tetrabutylammonium fluoride (TBAF) in THF used by Corey[14] is particularly effective causing removal of all of the silyl groups __16-20__ within 30 min. These conditions affect neither acid labile groups (such as the triphenylmethyl series) nor N-acyl groups. Phosphate protecting groups (cyanoethyl, CE; trichloroethyl, TCE; and phenyl) on phosphate triesters are not stable to TBAF in THF, being cleaved[17] within 30 min to the diester (22a). The fluoride ion catalyzed removal of phosphate protecting groups can be slowed by the addition of anhydrous acid (eg HOAc) permitting the removal of TBDMS without significant loss of the phosphate protecting group.[17,22] However, this is only a practical solution for short nucleotides.

The fluoride catalyzed removal of phenyl and trichloroethyl groups from phosphate triesters has led to a novel and efficient exchange reaction.[23,26] Thus, ϕ or TCE can be replaced by any 1°-alkyl group simply by dissolving the phosphate triester (eg 21b or c) in the appropriate alcohol (R"OH) along with cesium fluoride. Solid alcohols can be accommodated by using t-butanol as solvent which does not react. This result has led to a number of novel derivatives of nucleotides and other phosphate esters but a detailed discussion is beyond the objective of this report.

$$\text{21 a) } R = CE$$
$$\text{b) } R = TCE$$
$$\text{c) } R = \phi$$

$$\text{22 a) } R' = H$$
$$\text{b) } R'' = alkyl$$

I mentioned above that caution should be exercised when using TBAF in the presence of 0-acyl groups. I also mentioned the use of acetic acid to moderate fluoride ion catalysis. Both

of these points were nicely illustrated by an interesting observation by Baker.[27] He reported that treatment of the 5'-TBDMS-
2',3'-di-O-acylarabinoadenosine compounds 23 with anhydrous TBAF
often led to the 3',5'-di-O-acylarabinoadenosine (24) as the
major product. Addition of 1 eq. of acetic acid prevented this
rearrangement giving the expected 2',3'-di-O-acyl derivative 25.
Presumably in the absence of a ready proton source an initially
generated anion on the 5'-oxygen (23a) leads to 2'- to 5'-acyl
transfer.

One very useful manipulation that can be carried out with
silyl protecting groups is their ready replacement by acyl
groups.[28] For example, a TBDMS group in compound 26 can be
replaced by any acyl group present in an anhydride. When 26 is
stirred in THF with pivalic anhydride and TBAF for 10 h a 95%
yield of 27a is obtained. This is a remarkable result when
compared to the reaction between 5'-methoxytritylthymidine and
pivalic anhydride in pyridine which produces only 9% of 27a after
24 h. The TBDMS group can be replaced by a variety of acyl
groups giving for example 27b-d. It is not necessary to have a
TBDMS group present[28] since TBAF can catalyze the direct acylation of sterically hindered hydroxyls in THF. While we have to
date only investigated this reaction for nucleosides it should be

general. It has provided a very useful acylation procedure for nucleosides and avoids the problem of N-3 acylation of pyrimidine nucleosides.

There have been very few other silyl groups reported to have been used in nucleoside chemistry since our original reports. Hanessian[29] has described the use of the t-butyl diphenylsilyl group (TBDPS). This group was shown to be more stable than the TBDMS group toward acid. For example in compound 28 the R group (where R is trityl,tetrahydropyranyl or TBDMS) could be removed to give 28a (R=H) in 96% yield. Hata[29b] has used this group to

protect the O^6-position of guanosine and Engels[29c] applied the same group to the 5'-position of nucleosides.

Jones[30] reported in 1978 on the use of butyldiphenylsilyl groups in deoxynucleoside chemistry. The groups chosen were the

n-, sec-, and t-butyldiphenylsilyl groups of which the preferred
group was the t-butyldiphenylsilyl (TBDPS).   Jones et al. reacted
a number of deoxynucleoside 5'-monophosphates with TBDPS-Cl and
got derivatization on both the phosphate and 3'-hydroxyl groups
(30). The TBDPS group at the phosphate of 30 could be cleaved in

aqueous pyridine (12 h, RT) to give compounds of the type 31
which were used in the synthesis of an undecanucleotide.   The
3'-TBDPS group was only ~10% cleaved after 76 h in 9M NH$_4$OH at
25°C.   The 3'-sec-butyldiphenylsilyl groups was completely
cleaved in 4 h under these same conditions.   All three silyl
groups tested were cleaved with TBAF in pyridine in less than
20 h both from mononucleotide units 31 and from the 3'-end of the
undecanucleotide.   Jones et al. also showed that the lipophili-
city of the silyl groups made it possible to extract silylated
nucleotides such as 31 from aqueous solutions. The silyl groups
also produced improved retention of nucleotides on reverse phase
HPLC columns.

While the silyl groups in general and the t-butyldimethylsilyl (TBDMS) group in particular have been shown to be very
useful as protecting groups for deoxynucleosides, they have
provided a major advance in the protection of ribonucleosides.
The problems encountered in choosing protecting groups for ribonucleosides to be used in nucleotide synthesis were described at
the beginning of this report.   The successful procedures for
protecting ribonucleosides involved numerous steps which were
time-consuming and often of low yield.   Several authors[31] had

made contributions to the development of protecting groups in the
area. However, Nielson's synthesis[32] of a 2',5', N-protected
guanosine gives a realistic example of the pre-silyl groups
procedures.

We first reported the use of alkylsilyl groups with ribonucleosides in 1974.[33] Since that time we have extensively
explored[43-41] the use of the TBDMS groups in ribonucleoside and
ribonucleotide chemistry.

The degree of silylation of ribonucleosides is easily con-
trolled by the ratio of TBDMS-Cl to nucleoside.[35-36] Normally
1.2 eq. of TBDMS-Cl leads to 75-90% yields of the 5'-silylated
derivatives 32. By increasing the amount of silylating agent to
2-3 eq., the disilyl isomers 33 and 34 are obtained in yields of
40-50% and 30-40% respectively. A large excess 4-8 eq. of TBDMS-
Cl leads to the trisilyl derivatives 35 in >90% yields. Large

amounts of silylating agents must be used with guanosine at all levels.[36]   Reactions in DMF are normally complete within 2 h. The rate of reaction in pyridine is slower (up to 48 h) but selectivity toward the 2'-position is better.   Yields of 33 as high as 68% have been obtained in pyridine.   The 2'-isomers seem always to have higher R_f's than the 3'-isomer in all series studied to date.

The 2'- or 3'-silyl groups of 33 and 34 are more resistant to acid hydrolysis than the 5'-silyl group.   Consequently, the 2'- or 3'-monosilylnucleosides can be obtained in 60-75% yields on acid hydrolysis of 33 and 34.   The 2'- or 3'-silyl groups can be removed from 33 and 34 with 9M NH_4OH at 70°C within 4 h.   However, in the fully silylated derivative 34, the 2'- and 3'-silyl groups are completely stable under these same conditions.

For oligonucleotide synthesis in solution we prefer to use ribonucleosides bearing a 5'-methoxytrityl group.   Compounds 36 and 37 are readily obtained[35,36] in 40-65% yields from the appropriate 5'-methoxytritylribonucleoside.   Removal of the methoxy-

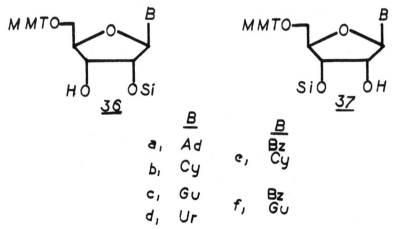

	B		B
a,	Ad	e,	Bz Cy
b,	Cy		
c,	Gu	f,	Bz Gu
d,	Ur		

trityl group gives the respective monosilyl derivatives in ~90% yields.   The preparation of 63 silylated derivatives including all possible mono, di, and tri TBDMS derivatives of A, C, U, G, G[Bz], C[Bz] and A[Bz] and their 5'-monomethoxytrityl derivatives has been fully described.[35,36] Pfleiderer has recently reported the synthesis of the same TBDMS derivatives of ribonucleosides.[42‾44]

Hernandez[45] has proposed the use of 4-dimethylaminopyridine (DAP) and triethylamine in DMF for the silylation of alcohols with TBDMS-Cl. However, his reported yield of 60% for the 5'-O-TBDMS uridine is lower than we routinely obtain (80%)[35] using standard silylation conditions in DMF or pyridine. One advantage proposed by Hernandez is the opportunity to use solvents other than DMF for silations. We investigated[46] the combination of DAP and triethylamine with TBDMS-Cl in DMSO, DMF, acetonitrile and dioxane for the silylation of adenosine. However, yields of 2',5'-disilyladenosine were all under 35% at temperatures from 20-50°C. The best yields were obtained using DAP in dioxane at 75°C (57%) or under milder conditions, imidazole in dioxane at 20°C (55%).

We have, however, found that soluble nitrates and perchlorates have a remarkable effect on the selectivity in the silylation of ribonucleosides.[46] When the reaction is carried out in THF with pyridine present, available 5' and 2'-positions are selectively silylated. In the absence of pyridine only 5'-silylation occurs. The active silylating agent appears to be the silylnitrate or silylperchlorate. Yields of the 2',5'-protected nucleosides are as high as 90% in the case of uridine. Interestingly, 4-acetylpyridine catalyzes a slower but highly selective reaction at the 2'-position. The anion is important since silver carbonate, silver trifluoroacetate, and silver carbonate have no effect on the silylation.

Perhaps the most remarkable catalytic activity is shown by DABCO (1,4-diazabicyclo[2,2,2]octane) in the presence of silver salts. In this case the silver ion is essential and the nature

$$HO \longrightarrow \underset{HO \quad OH}{\overset{O}{\bigcirc}} B \quad + \quad \underset{|}{\overset{|}{Si}} - Cl \quad \xrightarrow[\text{AgNO}_3 \text{ , THF}]{} \quad SiO \longrightarrow \underset{HO \quad OH}{\overset{O}{\bigcirc}} B$$

$$> 90\%$$

$$" \quad \xrightarrow[\substack{\text{AgNO}_3 \text{ , THF} \\ \text{pyridine}}]{"} \quad SiO \longrightarrow \underset{HO \quad OSi}{\overset{O}{\bigcirc}} B$$

B=U(90%) ;A(70%) ;C^{Bz}(82%)

G(60%) ;G^{Bz}(48%)

of the anion has no effect. Yields of 3',5'-diprotected ribonucleosides are as high as 90%. Only traces of 2'-silylation are observed. A high degree of selectivity toward 3'-silylation is also found when 4-nitropyridine N-oxide is used with silver salts. These selective silylations at the 3'-position appear to involve direct attack at the 3'-position since 2'-silylated derivatives appear to be stable under the reaction conditions.[47,48] This is a remarkably useful result and is apparently the first example of a monofunctional reagent showing this degree of selectivity for the 3'-position.

B=U	5%	91%
B=A	25%	70%
B=C^Bz	3%	93%
B=G^Bz	30%	60%

The procedures developed for the ribo series have been extended to the arabinonucleosides.[49,50] In the arabino series the DABCO conditions generate high yields of 2',5'-protected derivatives.

In the absence of DABCO but with silver salts and pyridine in THF silylation occurs at the 2'-position of araA but at the 3'-position of araC and araU. Replacing THF and pyridine with dimethoxyethane and triethylamine respectively reverses the isomer distribution and leads to 2'-silylation of araC and araU but to 3'-silylation of araA.

a) in THF,pyridine   major B=C,U   major B=A

b) in DME,Et$_3$N   major B=A   major B=C,U

Imbach[51,52] has recently developed an interesting route to silylated ribonucleosides. This route, involving coupling of a preformed sugar unit with a base, is suggested[51] as a route to unusual nucleosides where the base is other than the four common ones. Imbach has also suggested the use of potassium fluoride in the presence of dibenzo-18-crown-6 in THF for the removal of the silyl group.

$$B = \overset{Bz}{Ad} \text{ or } Ur; \; 93, 95\%$$

Recently Markiewicz has described[53,54] a very interesting bifunctional silylating agent, 1,3-dichloro-1,1,3,3-tetraiso-propyldisiloxane (TIPDS-Cl$_2$, 41). This reagent reacts with ribonucleosides to selectively protect the 3'- and 5'-positions to produce compounds 42 in 75% yields. Compounds of the type 42 will allow direct derivatization of the 2'-position. The silyl group can be completely removed from 42 using either TBAF in THF (10 min) or 0.2M HCl in dioxane-H$_2$O (4:1, 24 h). On the other hand 0.2M NaOH in dioxane-water (4:1, 25°C) or tri-n-butylammon-ium fluoride in THF (1M, 1 eq., 0.5 h) selectively cleave the

TIPDS group from the 3'-position to give $\underline{43}$ in 90% and 60% yields respectively.

Van Boom has employed the TIPDS group in preparing 2'-protected ribonucleosides.[55] He noted that care must be used with the TIPDS group since under certain acidic conditions (eg. in DMF) the group rearranged to the 2',3'-isomeric structure.

We have found that the tert-butyldimethylsilyl group (TBDMS) allows for considerable manipulation of nucleosides. For example the trisilylribonucleosides proved to be ideal intermediates in the preparation of N-levulinylnucleosides.[56] The levulinyl group is rapidly removed from either oxygen or nitrogen functions using hydrazine solutions. As such it is an excellent protecting group for certain situations. Levulinating the NH_2 groups of nucleosides was not an easy matter. TriTBDMScytidine and triTBDMSadenosine were smoothly converted to their N-levulinated derivatives using levulinic acid in THF with N-ethoxycarbonyl-2-ethoxy-1,2-

dihydroquinoline (EEDQ) in 95 and 80% yields respectively. TriTBDMSguanosine could only be N-levulinated by treatment first with butyllithium followed by addition of the pentafluorophenyl ester of levulinic acid for an overall yield of 20%. The TBDMS groups were smoothly removed using TBAF in THF to leave the N-levulinated nucleosides.

Leonard[57a] has recently used the TBDMS group for the protection of ribonucleotides during preparation of N-(dimethylamino)-methylene derivatives. Sung[57b] has also used TBDMS protection of sugar hydroxyls to allow ring modifications of thymidine.

The ability to prepare 2',5'- or 3',5'-diprotected ribonucleosides in highly selective reactions [47,48] has allowed for the easy manipulation of the 3'- or 2'-positions of the ribose ring. For example, the 2',5'-protected derivatives react with (thiocarbonyl)diimidazole followed by reduction with tributyltinhydride to give the 3'-deoxynucleosides.[58]

It is interesting that phenyl chlorothionocarbonate reacted readily with 2',5'-protected uridine and adenosine but not with the N-protected derivatives of A, C and G.[58] Robins[59] and Watanabe[60] used the Markeiwicz reagent (TLPDS) to protect the 3'-and 5'-positions during a similar preparation of 2'-deoxy-nucleosides.

We have recently[61] carried out the oxidation and reduction at the 2'- and 3'-positions of TBDMS-protected ribonucleosides. The reduction of the keto group leads to a predominance of the arabino product (from 2'-keto) or the xylo product (from 3'-keto).

There remain two important questions concerning the suita-bility of the silyl protecting groups for ribonucleotide synthe-sis. These are: (1) are the groups stable to the conditions used in phosphorylative coupling of nucleosides, and (2) can the silyl groups be removed successfully at the end of the condensa-tion procedures.

We have studied the stability of 2'-silylated nucleosides in a variety of solvents.[35,36,40] The 2'- (or 3'-) TBDMS group (in for example 33d) is completely stable in pure, dry pyridine, DMF and THF under conditions used for phosphorylation. We have synthesized nucleotides from nucleosides bearing a 2'-TBDMS group using 10 different condensing agents[41a] and in all cases the TBDMS group remained at the 2'-position. In general, one can state that the silyl group remains fixed at the 2'- (or 3'-) position in dry aprotic solvents. Results for the stability of 33d in several solvents are summarized below (as determined by HPLC). We have also compared six phosphate protecting groups in the phosphite coupling procedure[41b] and prepared all of the mixed ester units normally used in the traditional triester approach.[41a]

33d

Solvent	% remaining after 24 h
$CHCl_3$	100
DMF	100
CH_3CN	99.6
THF	99.3
pyridine	96.9
DMSO	95.6
DMF + imidazole (2 eq.)	80
CH_3CH_2OH	73.3
CH_3OH	64

While the TBDMS group remains stationary in dry aprotic solvents, it will isomerize between the 2'- and 3'-positions in protic solvents (eg ROH) or in basic media such as pyridine-water.[35,36,40] Isomerization is rapid in methanol being 14.5 times faster than in ethanol. The rate of isomerization depends

on the nucleoside, being generally fastest for uridine and adenosine. The ratios at equilibrium are summarized below for a
variety of nucleosides. The 2'-isomer generally predominates at
equilibrium (eg 64% 33a to 36% 34a).

Isomer Ratios at Equilibrium for Silylated Nucleosides
in Methanol at 30°C

% of Mixture at Equilibrium

isomer	uridine	cytidine	adenosine	guanosine	N^{Bz}-guanosine
2'-Si	57	60	57	—	52
3'-Si	43	40	43	—	48
5',2'-DiSi	64	—	56	58	—
5',3'-DiSi	36	—	44	42	—
5'-MMT-2'-Si	56	51	50	—	57
5'-MMT-3'-Si	44	49	50	—	43

We have so far not observed any isomerization in acidic
media.[35,36] While isomerization will occur in wet DMSO or in
pyridine-water, it can be prevented by the addition of sulfonyl
chlorides or sulfonic acids.[35] Isomerization does occur however
if silylated nucleosides are left in contact with silica gel for
extended periods.[35]

The above observations concerning isomerization of the TBDMS
groups are supported by the independent results obtained by
Pfleiderer[44] for N-benzoylcytidine derivatives. Pfleiderer[44]
also reported that isomerization was virtually non-existent in
pure anhydrous solvents. He also noted that the presence of
small amounts of acetic acid had a strong inhibitory effect on
isomerization. Recently Reese[62] has reported the synthesis of
2'-TBDMS-5'-acetyladenosine and he observed the same trend as
described above for the stability of the 2'-TBDMS group in a
similar range of solvents.

All of the results obtained to date show that while migration of the silyl groups can occur, it need not pose a threat to
nucleotide condensations.

We have made one additional observation that at first glance
is surprising.[36,40]   The N-benzoyl derivatives of adenosine and
cytidine that contain a silyl group on the 2'- and/or the 5'-
position undergo debenzoylation in methanol.

where   R = TBDMS   R' = H
                H   TBDMS
              MMT   TBDMS
             TBDMS   TBDMS

For N-benzoylcytidine derivatives the rate of debenzoylation
is much faster than isomerization (2'- to 3'-) and is easily
followed by HPLC.   The results are summarized below.

Debenzoylation of N-Benzoylcytidines in Methanol at 30°C

isomer	% Debenzoylation after 24 h	Time to complete loss of benzoyl
5'-SiC^{Bz}	49	~48 h
2'-SiC^{Bz}	34.5 (19.1% 2'; 15.4% 3')	~72 h
3'-SiC^{Bz}	7 (3.7% 2'; 3.3% 3')	23 days 50/50
5',3'-DiSiC^{Bz}	94.4 (85.7 2'; 8.7 3')	~30 h
5',3'-DiSiC^{Bz}	99 (30.4 2'; 68.6 3')	~24 h
5'-MMT-2'-SiC^{Bz}	98.4 (80.7 2'; 17.7 3')	~48 h
5'-MMT-3'-SiC^{Bz}	2.3 (0.8 2'; 1.5 3')	7 days 50/50

The N-benzoyladenosine derivatives undergo isomerization more rapidly than debenzoylation. For example, when 2',5'-di-TBDMS-N-benzoyladenosine (33e) is dissolved in methanol (7 h) at 30°C, the solution contains a mixture of 33e (77.7%), 3',5'-di-TBDMS-N-benzoyladenosine (34e, 17.0%), 2',5'-diTBDMS adenosine (33a, 2.5%) and its 3',5'-isomer, 34a (0.8%). Debenzoylation of the related guanosine derivatives essentially does not occur under these conditions.

The fact that debenzoylation only occurs when there is a TBDMS group on th 2'- and/or 5'-position, but does not occur for 3'-monoTBDMS derivatives, indicates a specific interaction between solvent, TBDMS and base ring. The rate of debenzoylation in ethanol is very slow. Only 4% of 33f is debenzoylated after 48 h in ethanol at 30°C.

NMR of Silated Nucleosides. We have looked at the ^{1}H and ^{13}C spectra of the silylated nucleosides.[35,36,63] While there are differences in the H$^{1'}$ chemical shifts and between the CH$_3$- and (CH$_3$)$_3$C-protons in isomeric silylated nucleosides, the most significant and useful shifts occur in the ^{13}C spectra. Silylation at a sugar hydroxyl leads to a downfield shift of the sugar carbon to which it is attached. This is true for all of the silylated ribonucleosides and arabinonucleosides.

Ribose ^{13}C Shifts* Relative to C-5' (ppm)

	C-1'	C-4'	C-2'	C-3'	C-5'0
adenosine	27.79	24.71	12.65	9.31	0.00
5'-SiA	26.34	22.06	12.44	7.46	0.00
2',5'-SiA	26.38	22.06	14.10	7.42	0.00
3',5'-SiA	26.27	22.66	11.22	9.66	0.00
cytidine	30.91	24.30	15.10	9.10	0.00
5'-SiC	29.21	22.10	13.94	7.16	0.00
2',5'-SiC	28.83	21.70	15.65	7.02	0.00
3',5'-SiC	29.80	22.29	13.82	8.92	0.00
5'-SiG	25.00	21.87	12.39	7.38	0.00
2',5'-SiG	24.98	21.85	14.36	7.44	0.00
3',5'-SiG	24.09	21.64	10.67	8.66	0.00
uridine	28.59	24.45	14.23	9.37	0.00
5'-SiU	27.23	22.14	12.92	7.47	0.00
2',5'-SiU	27.00	22.20	14.23	7.42	0.00
3',5'-SiU	27.09	22.56	12.37	9.37	0.00

*In pyridine-d_5. Data collected in collaboration with F. Hruska and W. G. Niemczura.

Some time ago Reese[64] observed that in the proton NMR of a pair of 2'- and 3'-isomers (position of substituent) the anomeric proton is at lower field (up to 0.25 ppm) for the 2'-isomer relative to the 3'-isomer. In addition $J_{1', 2'}$ is greater for the 3'-isomer than for the 2'-isomer. These rules were based on a series of substituted ribonucleosides where the substituent had either an inductive effect or through space shielding effect. It is therefore not surprising that the position of a TBDMS group, which would be expected to have very weak inductive or aniso-

tropic effects, is found to have very little effect on the anomeric (H-1') proton. The isomeric silylated nucleosides do not
conform to the general rules. For most (but not all) pairs
$J_{1',2'}$ is greater for the 3'-isomer than the 2'-isomer. Guanosine and adenosine derivatives generally have the anomeric proton
at greater chemical shift for the 2'- than 3'-isomer while C and
U derivatives generally show the reverse.

In pyrimidine (U,C) nucleosides the protons on the silicon
generally have very similar chemical shifts whether at the 2'-
position or at the 3'-position in silylated nucleosides. However, if there is a methoxytrityl group at the 5'-position, the
methyl and t-butyl protons are significantly shielded in the
3'-TBDMS derivative relative to the 2'-TBDMS derivative. For
purine (G,A) nucleosides the trend in silylated nucleosides is
quite different. For purines the methyl and t-butyl protons on a
TBDMS group in the 2'-position are significantly shielded relative to those in the 3'-isomer even in the case of the 5'-
tritylated derivatives.

Chemical Shifts of Anomeric and TBDMS Protons of Silylated
(2'- and 3'-isomers) Ribonucleosides*

Compound	H-1'($J_{1',2'}$,Hz)	$(CH_3)_3C$	$Si-CH_3$
2'-SiU	5.84(4.9)	0.95(9H)	0.17(6H)
3'-SiU	5.84(3.2)	0.95(9H)	0.15(6H)
2',5'-DiSiU	5.85(3.9)	0.97(9H);	0.17(6H);
		0.91(9H)	0.12(6H)
3',5'-DiSiU	5.86(4.2)	0.96(9H);	0.18(6H);
		0.95(9H)	0.16(6H)
5'-MMT-2'-SiU*	5.83(~3)	0.96(9H)	0.17(6H)
5'-MMT-3'-SiU	5.85(~3)	0.86(9H)	0.12(3H);
			0.05(3H)
2'-SiC^{Bz}	5.88(2.1)	0.95(9H)	0.20(3H);
			0.15(3H)
3'-SiC^{Bz}	5.91(2.4)	0.94(9H)	0.17(6H)

Compound	H-1'($J_{1',2'}$,Hz)	$(CH_3)_3C$	Si-CH$_3$
2',5'-DiSiC^{Bz}	5.90(1.5)	1.00(9H); 0.96(9H)	0.23(3H); 0.21(3H); 0.20(3H); 0.17(3H)
3',5'-DiSiC^{Bz}	5.94(2.9)	1.00(9H); 0.95(9H)	0.20(3H); 0.17(9H)
5'-MMT-2'-SiC^{Bz}	5.88(0.1)	0.97(9H)	0.27(3H); 0.20(3H)
5'-MMT-3'-SiC^{Bz}	5.97(1.8)	0.83(9H)	0.10(3H); 0.01(3H)
5'-MMT-2'-SiC	5.90(1.8)	0.95(9H)	0.22(3H); 0.17(3H)
5'-MMT-3'-SiC	5.92(2.5)	0.84(9H)	0.09(3H); -0.01(3H)
2'-SiG^{Bz*}†	5.87(6.0)	0.79(9H)	0.01(3H); -0.10(3H)
3'-SiG^{Bz*}†	5.82(6.0)	0.93(9H)	0.14(6H)
2',5'-DiSiG^{Bz*}	5.90(5.3)	0.95(9H); 0.81(9H)	0.14(6H); -0.02(3H); -0.12(3H)
3',5'-DiSiG^{Bz*}	5.86(4.5)	0.97(9H); 0.93(9H)	0.21(6H); -0.12(6H)
5'-MMT-2'-SiG^{Bz}	5.97(5.9)	0.84(9H)	0.06(3H); -0.07(3H)
5'-MMT-3'-SiG^{Bz}	5.95(6.2)	0.89(9H)	0.14(3H); 0.08(3H)
2'-SiG*†	5.68(6.0)	0.77(9H)	-0.03(3H); -0.12(3H)
3'-SiG*†	5.60(6.5)	0.91(9H)	0.11(6H)
2',5'-DiSiG*	5.84(5.0)	0.98(9H); 0.93(9H)	0.14(6H); 0.06(3H); 0.01(3H); 0.10(6H)
3',5'-DiSiG*	5.72(6.5)	0.93(9H); 0.75(9H)	0.22(6H)
5'-MMT-2'-SiG*	5.92(4.0)	0.87(9H)	0.08(3H); 0.00(3H)
5'-MMT-3'-SiG*	5.87(4.5)	0.90(9H)	0.17(3H); 0.09(3H)
2'-SiA	5.95(7.5)	0.78(9H)	-0.13(3H); -0.29(3H)
3'-SiA	5.95(7.5)	0.97(9H)	0.26(6H)
2',5'-DiSiA	6.07(4.5)	0.84(9H); 0.96(9H)	-0.08(3H); 0.01(3H); 0.15(6H)
5'-MMT-2'-SiA*	6.11(5.0)	0.85(9H)	-0.07(3H); 0.03(3H)

Compound	H-1'($J_{1',2'}$,Hz)	$(CH_3)_3C$	$Si-CH_3$
5'-MMT-3'-SiA*	6.09(4.0)	0.90(9H)	0.19(3H); 0.15(3H)

*Spectra were recorded at 90 MHz unless labelled with a * indicating 60 MHz spectra.

†Solvent was CD_3COCD_3 unless labelled by † indicating DMSO-d_6.

Nucleotide Synthesis. Our objective in introducing the alkylsilyl protecting groups and TBDMS in particular into the nucleoside area was to facilitate the synthesis of oligoribonucleotides. Therefore to fully establish the utility of these groups we must report on their compatibility with the existing technology for the formation of phosphate esters.

We have described in considerable detail[34-39] the fundamentals of an approach to oligonucleotide synthesis that combines the silyl protecting groups with the phosphorodichloridite condensation procedure which was originally introduced by Letsinger.[65] These two tools are remarkably compatible. Silylating agents do not react with amino groups in nucleosides; neither do phosphorodichloridites. This results in a considerable reduction in the number of steps necessary for the protection of ribonucleosides. It also has the advantage that depurination of adenosine units is avoided during acidic removal of methoxytrityl or other acid labile groups. The glycoside bond in N-acylated adenosine units is sensitive to acid hydrolysis.[66]

The general approach to oligoribonucleotide synthesis is outlined below for the phosphorodichloridite procedure. The procedure requires far less time than any other published procedure. The first step requires <15 min and the second step

<30 min in a one-pot procedure. The iodine oxidation step re-
quires <5 min. Yields for all nucleoside combinations are in the
60-85% range. Our initial published results[35,36] described
longer reaction times but we have found this to be unnecessary.
For the purpose of maximizing the TLC mobility of the nucleotide
products such as 5, we generally use N-protection (benzoyl) on
cytidine and guanosine (eg. 36e and 36f).

The final aspect of the overall strategy for the synthesis
of long chain nucleotides lies with the choice of R in 4 (and 5)
bearing in mind the requirements set out at the beginning of this
report. The R that we have chosen is the levulinyl group
($CH_3COCH_2CH_2CO-$, Lv) first introduced by Guthrie and Lucas[67] to
carbohydrates and extended to nucleosides by Hassner and
Patchornik.[68] Van Boom[69] has described the utility of the levulinyl group for 5'-protection in the procedures his group has
developed for oligonucleotide synthesis.

The levulinyl group is easily introduced into our system via
compound 36. The direct levulination of 36 occurs in >90% yields
for all bases.[39] No isomerization of the silyl group occurs in
36 during the production of 44. Acid converts 44 to 45 (4 where
R'''=Lv).

The key dinucleotide unit 3 can now be fully described as 46 and allows for chain extension in either direction since treatment with acid (acetic or toluenesulfonic acid) removes the methoxytrityl group to give 47 while treatment with hydrazine (5 min) removes only the levulinyl group to give 48. The strategy is now complete since units 47 and 48 allow for block condensations.

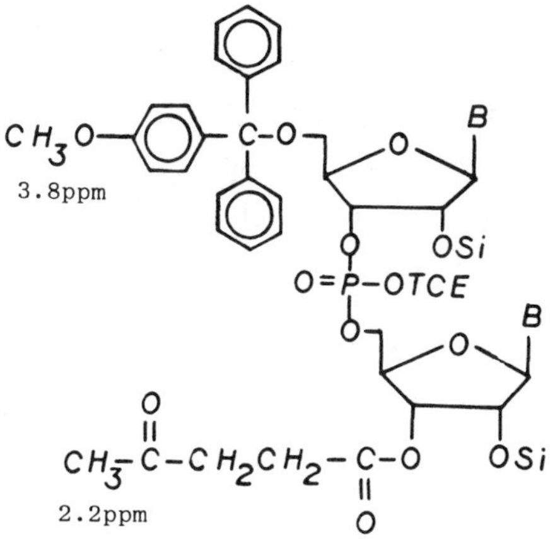

46

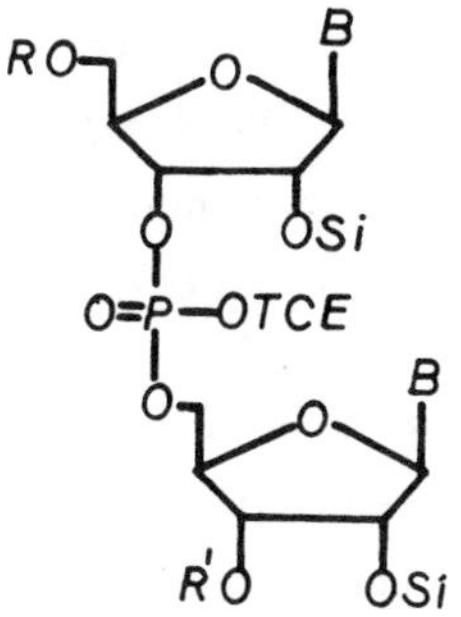

47, R=H, R′=Lv
48, R=MMT, R′=H

The synthesis of oligoribonucleotides is illustrated by the synthesis of a hexadecauridylic acid by a block condensation procedure as illustrated below.

MMT-U_P^{Si}(TCE)U_{OH}^{Si} + HOU_P^{Si}(TCE)U_{Lv}^{Si}

48 (B=Ur) 47 (B=Ur)

↓

MMT-U_P^{Si}(TCE)[U_P^{Si}(TCE)]$_2U_{Lv}^{Si}$

49

↙ ↘

MMT-U_P^{Si}(TCE)[U_P^{Si}(TCE)]$_2U_{OH}^{Si}$ HOU_P^{Si}(TCE)[U_P^{Si}(TCE)]$_2U_{Lv}^{Si}$

50 51

↓

MMT-U_P^{Si}(TCE)[U_P^{Si}(TCE)]$_6U_{Lv}^{Si}$

52

↙ ↘

MMT-U_P^{Si}(TCE)[U_P^{Si}(TCE)]$_6U_{OH}^{Si}$ HOU_P^{Si}(TCE)[U_P^{Si}(TCE)]$_6U_{Lv}^{Si}$

53 54

↓

MMT-U_P^{Si}(TCE)[U_P^{Si}(TCE)]$_{14}U_{Lv}^{Si}$

55

All of the condensation steps leading up to 55 were carried out using a ratio of trityl block to untritylated block of 1:0.8 and reaction times of 15 min in the first step and 30 min in the second step. Yields ranged from an average of 75% for 46 and 73% for 52 to 51% for 55. One very important characteristic of the protected products from these condensation reactions is that the desired products (eg. 49, 52, 55) all have a methoxy (CH_3O-)

$$O$$

group at the 5'-end of the chain and a CCH_3 group (from LV) at the 3'-end of the chain. Consequently only the desired product shows two clean singlets in the NMR at approximately 3.85 ppm

$$O$$

(methoxy) and 2.26 ppm (CH_3C-).

The TBDMS protection of ribonucleosides also allowed us to prepare a heptaribonucleotide[37] in a stepwise fashion. This sequence, corresponding to the terminal sequence of tRNA$_{E. coli}^{fmet}$ contained A, C and G units. The protected oligonucleotides produced during this synthesis were subjected to plasma desorption mass spectrometry.[70] Excellent spectra were obtained for these compounds and fragmentation patterns allowed for sequence determination.

$$\text{MMT-G}\ ^{Bz}_{\ P}{}^{Si}(\text{TCE})\text{C}\ ^{Bz}_{\ P}{}^{Si}(\text{TCE})\text{A}^{Si}_{\ P}(\text{TCE})\text{A}^{Si}_{\ P}(\text{TCE})\text{A}^{Si}_{\ P}(\text{TCE})\text{C}\ ^{Bz}_{\ P}{}^{Si}(\text{TCE})$$

$$\text{C}\ ^{Bz}_{\ P}{}^{Si}(\text{TCE})\text{A}^{Si}_{Si}$$

One must also mention the 3',5'-protected nucleosides which to this point have almost appeared to be useless side products. In fact they are very valuable side products. With the discovery[1-3] that the 2',5'-linked trinucleotide of adenosine is a key molecule in the interferon process, the ability to synthesize such compounds is a necessity. We prepared[37] this essential molecule using 5'-methoxytrityl-3'-TBDMS adenosine (37a) and 2',3'-di-TBDMS adenosine, preparing first the dinucleotide and extending it from the 5'-end to give the fully protected nucleotide 56. Yields using the phosphorodichloridite procedure were 68% at the dinucleotide level and 66% for the tri.

The deprotection of 56 to produce the free nucleotide 57 illustrates the removal of the protecting groups from nucleotides in general. The first step involves removal of the methoxytrityl group with acetic acid (or benzenesulfonic acid). The second step involves treatment with the zinc-copper couple in DMF at 50°C for 2-3 h and finally removal of the silyl groups with TBAF in THF for 30 min. It is essential to remove the phosphate protecting group before treatment with TBAF to prevent internucleotide bond cleavage.[71]

The 2'-5'-linked nucleotides such as 57 are degraded by snake venom but not by spleen phosphodiesterase. We have pre-

pared compound 58 (which is isomeric to 44) for all the common
bases and a study of 2',5'-linked nucleotides is under way.

58

Several recent reports from other laboratories have described the synthesis of ribonucleotides using ribonucleosides
protected by the TBDMS group.[53,72-74]

Protecting grops having silicon β to the protected site have
also been used to protect nucleosides and nucleotides. Hata[75]
has described the use of the 2-diphenylmethylsilylethyl group as
a phosphate protecting group in the triester synthesis of nucleotides. The group is reported to be stable to mild acid and base
but easily removed using TBAF in acetonitrile. Chatopadhyaya[76]
used a similar side chain as part of an alkoxycarbonyl group to
protect the hydroxyl groups of nucleosides.

$$\begin{array}{cc} \phi & O \\ Me-Si-CH_2CH_2OP-OR & \\ \phi & OR' \end{array} \qquad \begin{array}{cc} Me & O \\ Me-Si-CH_2CH_2O-C-OR & \\ Me & \end{array}$$

Solid Support Synthesis of Ribonucleotides. We have used silylated ribonucleosides as the key reagents in the synthesis of
ribonucleotides on a solid support.[77,78] The procedures employ
the phosphodichloridate coupling procedures, silica as solid
support[79,80] and 2'-TBDMS ribonucleosides. These procedures have
led to the synthesis of a 19-unit ribonucleotide corresponding to
units 9-27 of the formylmethionine t-RNA from E. coli.

```
            C G                  9
      C            A            G(5')
                C G A G
   U            G C U C
      G         A            G(3')
         G U                  27
```

Solid Support Synthesis of Oligoribonucleotides

CONCLUSION

We believe that the TBDMS group has greatly simplified the problem of protecting ribonucleosides. These TBDMS-Protected ribonucleosides which are obtained rapidly are coupled together most rapidly by the phosphorodichloridite procedure. They are also compatible with all of the existing generally used coupling procedures. The TBDMS groups contribute some interesting features as well, such as the novel debenzoylations and the isomerations in methanol. While we have been most intent on pushing ahead rapidly to prove the potential of our methods for oligoribonucleotide synthesis, we hope shortly to provide the results of a detailed study on these compounds through their ^{1}H, ^{13}C, ^{31}P and ^{29}Si NMR spectra. Finally, we conclude that we have found a protecting group that is essentially stable to the conditions of acid and base used during nucleotide synthesis and which can be removed under "almost" neutral conditions. This group, the TBDMS group, has the added value of permitting characterization of the protected nucleosides by gas chromatography and mass spectrometry and allowing the sequencing of protected nucleotides by mass spectrometry.

REFERENCES

1. B. R. G. Williams and I. M. Kerr, Nature, 276, 88-90 (1978).
2. A. Schmidt, A. Silverstein, L. Shulman, P. Federman, H. Berissi and M. Revel, FEBS Letters, 95, 257-264 (1978).
3. I. M. Kerr and R. E. Brown, Proc. Natl. Acad. Sci. USA, 75, 256-260 (1978).
4. R. L. Letsinger and K. K. Ogilvie, J. Am. Chem. Soc., 89, 4801-4803 (1967); R. L. Letsinger and V. Mahadevan, ibid., 87, 3526-3531 (1965); 88, 5319-5324 (1966).
5   a. There are several variations in this approach but the protecting group problems are similar in all of them. b. C. B. Reese, Phosphorus and Sulfur, 1, 245-260 (1976) B. E. Griffin, M. Jarman and C. B. Reese, Tetrahedron, 24, 639-662 (1968); H. P. M. Fromageot, C. B. Reese and J. E. Sulston, ibid., 24, 3533-3540 (1968).
6. K. K. Ogilvie and D. J. Iwacha, Can. J. Chem., 52, 1787-1797 (1974); K. K. Ogilvie and L. A. Slotin, ibid., 51, 2397-2405 (1973).
7. S. Uesugi, J. Yano, E. Yano and M. Ikehara, J. Am. Chem. Soc., 99, 2313-2323 (1977); ibid., 96, 4966-4972 (1974).
8. K. L. Agarwal and M. M. Dhar, Tetrahedron Lett., 2451-2452 (1965); Y. Mizirno and T. Sasaki, ibid., 4579-4584 (1965).
9. S. N. Alam and R. H. Hall, Anal. Biochem., 40, 424-428 (1971).
10. H. Gilman and G. E. Dunn, Chem. Rev., 52, 77115 (1953), H. Gilman and R. N. Clark, J. Am. Chem. Soc., 69, 1499-1500 (1947); H. Gilman and C. G. Brannen, ibid., 73, 4640-4644 (1951).
11. L. H. Sommer and L. J. Taylor, J. Am. Chem. Soc., 76, 1030-1034 (1954); W. H. Nebergall and O. H. Johnson, ibid., 71, 4022-4024 (1949), A. D. Allen, J. C. Charlton, C. Eaborn and G. Modena, J. Chem. Soc., 3668-3670 (1957); A. D. Allen and G. Modena, ibid., 3671-3678 (1957).

12. G. Stork and P. F. Hudrilik, J. Am. Chem. Soc., $\underline{90}$, 4462-4464 (1968).

13. R. L. Hancock, J. Gas. Chromatog., $\underline{6}$, 431-438 (1968).

14. E. J. Corey and A. Venkateswarlu, J. Am. Chem. Soc., $\underline{94}$, 6190-6191 (1972).

15. K. K. Ogilvie and D. J. Iwacha, Tetrahedron. Lett. 317-319 (1973).

16. K. K. Ogilvie, Can. J. Chem., $\underline{51}$, 3799-3807 (1973).

17. K. K. Ogilvie, S. L. Beaucage, D. W. Entwistle, E. A. Thompson, M. A. Quilliam and J. B. Westmore, J. Carb. Nucs. Nuctds., $\underline{3}$, 197-227 (1976).

18. M. A. Quilliam, K. K. Ogilvie and J. B. Westmore, J. Chromatog., $\underline{105}$, 297-307 (1975).

19. M. A. Quilliam, K. K. Ogilvie and J. B. Westmore, Biomed. Mass. Spec. $\underline{1}$, 78-79 (1974).

20. a. M. A. Quilliam, K. K. Ogilvie, K. L. Sadana and J. B. Westmore, Org. Mass. Spec., $\underline{15}$, 207-219 (1980); b. M. A. Quilliam, K. K. Ogilvie and J. B. Westmore, ibid., $\underline{16}$, 129-138 (1981).

21. a. M. A. Quilliam, K. K. Ogilvie, K. L. Sadana and J. B. Westmore, J. Chromatog., $\underline{196}$, 367-378 (1980); b. $\underline{194}$, 379-386 (1980).

22. K. K. Ogilvie, S. L. Beaucage and D. W. Entwistle, Tetrahedron Lett., 1255-1256 (1976).

23. K. K. Ogilvie and S. L. Beaucage, J. Chem. Soc. Chem. Comm., 443-444 (1976).

24. K. K. Ogilvie, S. L. Beaucage, N. Theriault and D. W. Entwistle, J. Am. Chem. Soc., $\underline{99}$, 1277-1278 (1977).

25. K. K. Ogilvie, S. L. Beaucage, M. F. Gillen and D. W. Entwistle, Nuc. Acids. Res., $\underline{6}$, 2261-2274 (1979).

26. K. K. Ogilvie and S. L. Beaucage, ibid., $\underline{7}$, 805-824 (1979).

27. D. C. Baker, T. H. Haskell, S. R. Putt and B. J. Sloan, J. Med. Chem., $\underline{22}$, 273-279 (1979).

28. S. L. Beaucage and K. K. Ogilvie, Tetrahedron Lett., 1691-1694 (1977).

29. a. S. Hanessian and P. Lavallee, Can. J. Chem., 53, 2975-2977 (1975). b. H. P. Daskalou, M. Sekine and T. Hata, Tetrahedron Lett., 21, 3899-3902 (1980). c. J. Engels, ibid., 21, 4339-4342 (1980).

30. R. A. Jones, H.-J. Fritz and H. G. Khorana, Biochem., 17, 1268-1278 (1978).

31. H. Kossel and H. Seliger, Progress in the Chemistry of Organic Natural Products, pp. 297-508, W. Herz, H. Grisebach and G. W. Kirby editors, Springer-Verlag, New York, 1975.

32. T. Nelson, E. V. Wastrodowski, and E. S. Werstiuk, Can., J. Chem., 51, 1068-1074 (1972).

33. K. K. Ogilvie, K. L. Sadana, E. A. Thompson, M. A. Quilliam and J. B. Westmore, Tetrahedron. Lett., 2861-2863 (1974).

34. K. K. Ogilvie, N. Theriault and K. L. Sanana, J. Am. Chem. Soc., 99, 7741-7743 (1977).

35. K. K. Ogilvie, S. L. Beaucage, A. L. Schifman, N. Y. Theriault and K. L. Sadana, Can. J. Chem., 56, 2768-2780 (1978).

36. K. K. Ogilvie, A. L. Schifman and C. L. Penney, ibid., 57, 2230-2238 (1979).

37. K. K. Ogilvie and N. Y. Theriault, Tetrahedron Lett., 2111-2114 (1979).

38. K. K. Ogilvie and N. Y. Theriault, Can. J. Chem., 57, 3140-3144 (1979).

39. K. K. Ogilvie and M. J. Nemer, Can. J. Chem., 58, 1389-1397 (1980).

40. K. K. Ogilvie and D. W. Entwistle, Carbohyd. Res., 89, 203-210 (1981).

41. a. K. K. Ogilvie and R. N. Pon, Nuc. Acids Res., 8, 2105-2115 (1980). b. K. K. Ogilvie, N. Y. Theriault, J. M. Seifert, R. T. Pon and M. J. Nemer, Can. J. Chem., 58, 2686-2693 (1980).

42. D. Flockerzi, G. Silber, R. Charubala, W. Schlosser, R. S. Varma, F. Creegan and W. Pfleiderer, Liebigs Ann. Chem., 1568-1585 (1981).

43. G. Silber, D. Flockerzi, R. S. Varma, R. Charubala, E. Uhlmann and W. Pfleiderer, Helv. Chim. Acta, 64, 1704-1716 (1981).

44. W. Kohler and W. Pfleiderer, Liebigs Ann. Chem., 1855-1871 (1979); W. Kohler, W. Scholosser, G. Charubala and W. Pfleiderer, Chemistry and Biology of Nucleosides and Nucleotides, pp. 347-348, R. E. Harmon, R. K. Robins and L. B. Townsead, Ed., Academic Press, New York, 1978.

45. S. K. Chaudary and O. Hernandez, Tetrahedron Lett., 99-102 (1979).

46. K. K. Ogilvie and R. Pon, unpublished results.

47. G. H. Hamkimelahi, Z. A. Proba and K. K. Ogilvie, Can. J. Chem., 60, 1106-1113 (1982); Tetrahedron Lett., 22, 4775-4778 (1981).

48. G. H. Hakimelahi, Z. A. Proba and K. K. Ogilvie, Tetrahedron Lett. 22, 5243-5246 (1981).

49. K. K. Ogilvie, G. H. Hakimelahi, Z A. Proba and D. P. C. McGee, ibid., 23, 1997-2000 (1980).

50. D. P. C. McGee, Master Thesis, McGill University, 1981.

51. F. Dumont, R H. Wightman, J. C. Ziegler, C. Chavis and J. L. Imbach, Tetrahedron Lett., 3291-3294 (1979).

52. C. Chavis, F. Dumont, R. H. Wightman, J. C. Ziegler and J. L. Imbach, J. Org. Chem., 47, 202-206 (1982).

53. W. T. Markiewicz, J. Chem. Res. (S), 24-25 (1979), W. T. Markiewicz and M. Wiewiorowski, Nuc. Acids Res., Special Pub. #4, 5185-5187 (1978).

54. W. T. Markiewicz, N. Sh. Padyukova, Z. Samek and J. Smrt., Coll. Czech. Chem. Comm., 45, 1860-1865 (1980).

55. C. H. M. Verdegall, P. L. Janssee, J. F. M. de Rooij, G. Veeneman and J. H. van Boom., Recl. Trav. Chim. Pays-Bas, 100, 200-204 (1981); Tetrahedron Lett., 21, 1571-1574 (1980).

56. K. K. Ogilvie, M. J. Nemer, G. H. Hakimelahi, Z. A. Proba and M. Lucas, Tetrahedron Lett., 23, 2615-2618 (1982).

57. a. A. W. Czarnik and N. J. Leonard, J. Am. Chem. Soc., 104, 2264-2631 (1982). b. W. L. Sung, J. C. S. Chem. Comm., 1089 (1981).

58. K. K. Ogilvie, G. H. Hakimelahi, Z. A. Proba and N. Usman, submitted for publication.

59. M. J. Robins and J. S. Wilson, J. Am. Chem. Soc., 103, 932-933 (1981).

60. K. Pankiewicz, A. Matsuda and K. A. Watanabe, J. Org. Chem., 47, 485-488 (1982).

61. J. Cormier and K. K. Ogilvie, unpublished results.

62. S. S. Jones and C. B. Reese, J. C. S. Perkin I, 2762-2764 (1979).

63. A. L. Schifman, Ph.D. Thesis, McGill University, 1980.

64. H. P. M. Fromageot, B. E. Griffin, C. B. Reese, J. E. Sulston and D. R. Trentham, Tetrahedron, 22, 705-710 (1966).

65. R. L. Letsinger and W. B. Lunsford, J. Am. Chem. Soc., 98, 3655-3661 (1976).

66. H. Schaller, G. Weimann, B. Lerch and H. G. Khorana, ibid., 85, 3821-3827 (1963).

67. D. Guthrie and T. J. Lucas, Carb. Res., 33, 391-395 (1974).

68. A. Hassner, G. Strand, M. Rubinstein and A. Patshornik, J. Am. Chem. Soc., 97, 1614-1615 (1975).

69. J. H. van Boom and P. M. J. Burgers, Rec. Trav. Chim. Pay-Bas, 97, 73-80 (1978), Tetrahedron Lett., 4875-4878 (1976).

70. C. J. McNeal, K. K. Ogilvie, N. Y. Theriault and M. J. Nemer, J. Am. Chem. Soc., 104, 981-984; 976-980; 972-975 (1982).

71. K. K. Ogilvie and S. L. Beaucage, Nuc. Acids Res., $\underline{7}$, 805-823 (1979).

72. J. A. J. den Hartog and J. H. van Boom, Synthesis, 599-600 (1979); J. A. J. den Hartog, R. A. Wijnands and J. H. van Boom, J. Org. Chem., $\underline{46}$, 2242-2251 (1981).

73. J. F. M. de Rooij, G. Wille-Hazeleger, P. M. J. Burgers and J. H. van Boom, Nuc. Acids Res., $\underline{6}$, 2237-2259 (1979).

74. K. L. Sadana and P. C. Loewen, Tetrahedron. Lett., 5095-5098 (1978).

75. S. Honda, T. Kamimura, M. Sato, K. Terado, M. Sekine and T. Hata, Nuc. Acids Res., Symp. Ser. #6, 5185-5186 (1979); S. Honda and T. Hata, Tetrahedron Lett., $\underline{22}$, 2093-2096 (1981).

76. C. Gioeli, N. Balgobin, S. Josephson and J. B. Chattopadhyaya, ibid., $\underline{22}$, 969-972 (1981).

77. K. K. Ogilvie and M. J. Nemer, ibid., 21, 4159-4162 (1980).

78. M. J. Nemer, Ph.D. Thesis, McGill University, 1982.

79. G. Alvarado-Urbina, G. M. Sathe, W.-C. Liu, M. F. Gillen, P. D. Duck, R. Bender and K. K. Ogilvie, Science, $\underline{214}$, 270-274 (1981).

80. M. D. Matteucci and M. H. Caruthers, J. Am. Chem. Soc., $\underline{103}$, 3185-3191 (1981).

UNUSUAL NUCLEOSIDE SYNTHONS
AND OLIGONUCLEOTIDE SYNTHESIS

Gilles Gosselin
Abd El Fattah Haïkal
Claude Chavis
Jean-Louis Imbach

Bio-Organic Laboratory
University of Sciences and Techniques of Languedoc
Montpellier, France
(Chemical Synthesis)

Deborah A. Eppstein
Y. Vivienne Marsh
Brian B. Schryver

Institute of Bio-Organic Chemistry
Syntex Research
Palo Alto, California
(Biological Experiments)

I. INTRODUCTION

Oligonucleotide synthesis is well documented but limitations exist such as obtaining sufficient quantities of starting nucleosidic synthons which are dissymmetrically substituted with suitable

Contribution number 161 from the Institute of Bio-Organic Chemistry, Syntex.

protecting groups (1-3). Protected deoxynucleosides of natural bases are now commercially available and thus the synthesis of some DNA fragments can be easily performed. However, this is not the case for ribonucleosides and also for unusual and/or modified nucleosides, for which one has to perform a multi-step synthesis before reaching the starting synthons. The usual method to obtain these synthons is to first synthesize the nucleoside (except for ribonucleosides which are commercially available), and then to introduce on the sugar moiety the desired substituent or protecting group.

This type of approach is quite time consuming and necessitates a differentiation between the three hydroxyl functions which leads to isomeric mixtures and hence separation problems (4-6). It is possible to avoid such problems by preparing a dissymmetrically substituted sugar prior to the glycosylation procedure.

Therefore we would like to focus on the following topics :
1) the synthesis of some 2'-O substituted ribofuranonucleosides and
2) the preparation of some specific xylofuranonucleosides and the formation of the corresponding $(XyloA2'p)_2XyloA$.

II. 2'-O-SUBSTITUTED RIBOFURANONUCLEOSIDES

Concerning the first point, we were interested in obtaining some $2'-OCH_3$ and 2'-OTBDMS derivatives of U and suitably amino protected A, G and C. The $2'-OCH_3$ nucleosides are among the rare constituents of t-RNA (7) and the O-TBDMS protecting group is currently used in oligonucleotide synthesis as it is stable in mildly acidic or non-acidic conditions (5,8). As mentioned previously, past work which utilized such 2'-substituted intermediates for oligonucleotide synthesis involved their preparation at the nucleoside stage and gave rise to isomeric mixtures whose separation were, at times, extremely difficult (9,10). Therefore, our goal was to introduce 2',3'-differentiation at the carbohydrate stage while taking into account the regioselectivity and stereospecificity of the glycosyl coupling reaction.

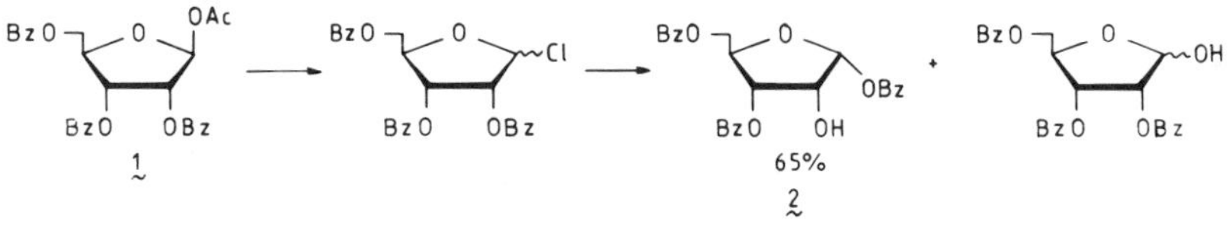

_ Scheme I _

Thus, as starting material, we needed an easily accessible
2'-OH ribofuranose derivative and we focused our attention on
1,3,5-tri-O-benzoyl-α-D-ribofuranose 2 described by Fletcher more
than twenty years ago (11). This compound was of interest to us as
various substituents could be introduced on position 2 of the
sugar and furthermore it retains an acyl leaving group on the anomeric
position required for glycosylation. The α configuration does
not preclude nucleosidic condensation as we have previously reported from our laboratory (12).

Therefore, 1-O-acetyl-2,3,5-tri-O-benzoyl-β-D-ribofuranose
(1, commercially available) was converted to the 1-chloro derivative which on hydrolysis gave a mixture of the corresponding

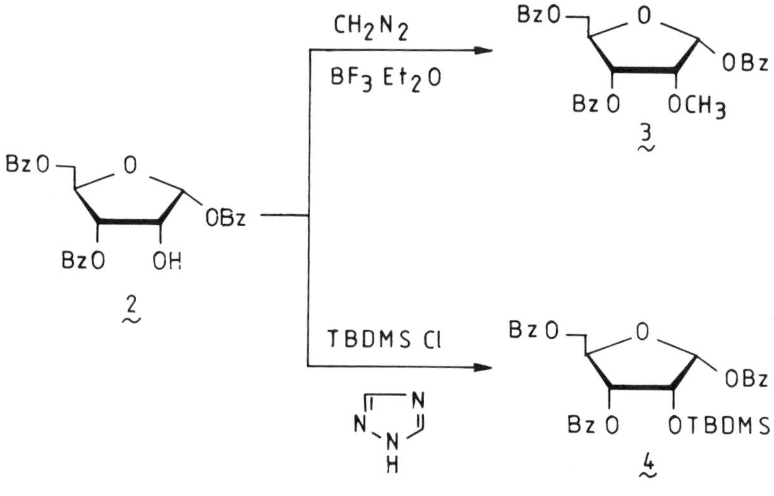

_ Scheme II _

2-hydroxy and 1-hydroxy sugars. However, 2 is easily separated as
it crystallizes directly from the solution with a 65 % yield.
Reaction of 2 with either diazomethane and BF_3/Et_2O as catalyst
or t-butyldimethylsilylchloride (TBDMSCl) and 1,2,4-triazole as
catalyst gave the desired sugars 3 and 4 with yields of 75 % and
100 % respectively (13).

Next we turned our attention to the nucleosidic condensation
using these two starting sugars with the appropriately protected
purines and pyrimidines. Tables I and II describe the results of
our condensation reactions. The yield represented for each nucleoside analog is obtained after chromatographic purification. All
these compounds are of β configuration and their structures were
established as N-9 for the purines and N-1 for the pyrimidines on
the basis of their UV and ^{13}C NMR spectra. It is worth noting,
that the use of such non-participating substituents at position 2
of the sugar does not prevent regioselectivity and stereospecificity during glycosylation.

Subsequent removal of the 3' and 5' benzoyl groups with
sodium methoxide gave high yields of the 2'-OCH_3 synthons which then

TABLE I   *Condensation yields with 2-OMe sugar*

$B-X$	METHOD	ISOLATED YIELD
C^{iBu}	TMS derivative CF_3SO_2TMS catalyst	55 %
A^{Bz}	HgCl derivative $TiCl_4$ catalyst	56 %
A^{Me_2}	$SnCl_4$ catalyst	35 %
C^{Bz}	$SnCl_4$ catalyst	70 %
U	$SnCl_4$ catalyst	75 %

+ 25 % N-7 β

TABLE II. Condensation yields
with 2-OTBDMS sugar

$B^{\nearrow X}$	Method	Isolated Yield
G^{iBu}	CF_3SO_2TMS	36%[+]
A^{Bz}	HgCl derivative $TiCl_4$ catalyst	93%
A^{Me}_2	$SnCl_4$ catalyst	36%
C^{Bz}	$SnCl_4$ catalyst	86%
U	$SnCl_4$ catalyst	95%

[+]16% N-7β

$$B^{\nearrow X} + \underset{\underset{\underset{4}{\sim}}{BzO\quad OTBDMS}}{\overset{BzO}{\diagdown}\underset{OBz}{\bigcirc}} \longrightarrow$$

can be used directly for oligonucleotide synthesis. Deprotection
of the corresponding 2'-OTBDMS nucleosides, however, is more complicated because base-catalyzed debenzoylation with sodium methoxide causes isomerisation of the DBSMS group (8,14). Similarly,
elimination of the silyl group with fluoride ion led to a mixture
of the 3',5' and 2',5' di-0-benzoyl derivatives (15). All of the
procedures we used in an attempt to avoid such isomerisation pro-
cesses, i.e., either sodium borohydride in ethanedithiol for the
removal of the benzoyl groups and antimony pentafluoride or acidic
conditions for deprotection of the TBDMS function, failed and gave
rise in each case to isomeric mixtures (16). Although we obtained
mixtures, they could be easily separated and furnished the desired
analog as the major constituent.

III. SUBSTITUTED XYLOFURANOSYLADENINES
AND (XYLOA2'p)$_2$XYLOA

The second aspect of our presentation is connected with the
preparation of (xyloA2'p)$_2$xyloA 5 (17). The rationale for the
synthesis of this unusual oligonucleotide is that the natural

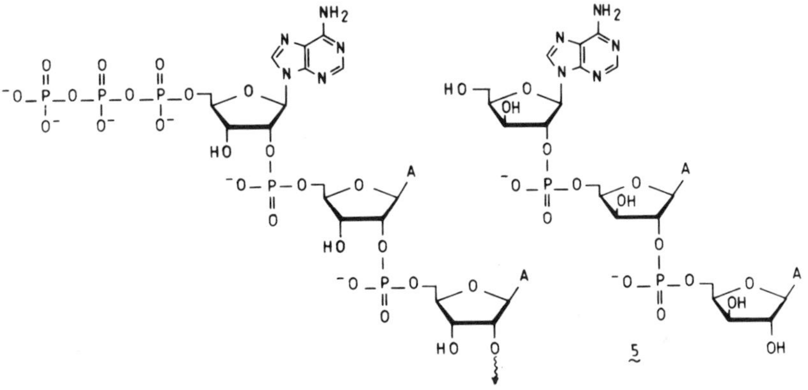

Natural 2′—→5′ oligo A,
" mediators " of interferon

Trimeric xylo analogue
of the natural 2′—→5′
oligo A core

Figure 1

species $ppp(A2'p)_nA, n \geqslant 2$ (18) (Fig. 1) has been suggested to play
a role in the interferon-mediated establishment of the antiviral
state (19-26) as well as to direct cell growth inhibition caused
by interferons (26-31). These $ppp(A2'p)_nA$, which are produced by
an interferon-induced oligoadenylate synthetase, activate a latent
endogenous endonuclease (32-54) ; however they are rapidly degra-
dated from the 3'(2') end by a cellular phosphodiesterase (28,32,
55-56). This has aroused interest in the potential use of these
2'→5' oligomers or analogs to directly achieve antiviral and/or
antitumor effects without use of interferons per se. Since these
compounds are highly negatively charged molecules, the $ppp(A2'p)_nA$
normally do not enter intact cells, and thus artificial means such
as hypertonic shock (20-21) or calcium phosphate co-precipitation
(23,26,57-58) are necessary for the introduction of these oligomers into cells. Such techniques are applicable only for in vitro
experiments but not for in vivo use. Thus it is preferable to use
less-highly charged species to facilitate entry into intact cells.
Among these species, the simplest is the trimeric core $(A2'p)_2A$
which was chemically or enzymatically synthesized by various proce-

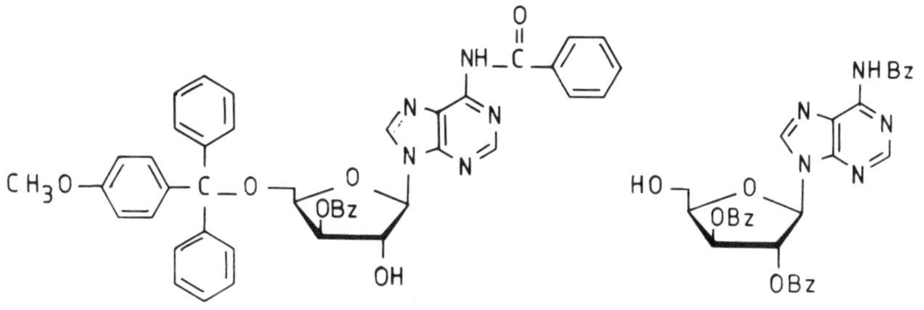

First Synthon : 6   second Synthon : 7

_ Figure 2 _

dures (59-77) and which, despite the fact that it does not bind
efficiently to the endonuclease (78) and is inactive in cell-free
extracts (79), inhibits DNA synthesis in certain cells (20,27-29,
31,80-81). Another therapeutic factor to be considered for these
oligomers is to decrease their intracellular degradation by the
phosphodiesterase. Along these lines, different analogs of the
natural core have been recently synthesized, with modifications on
the sugar moiety (69,74,82-86) or on the internucleotidic phosphate linkage (87). Some of these compounds exhibited important biological properties (80,84,88). For our approach, we anticipated
that inversion of configuration at 3', i.e., xylo instead of ribo,
would lead to an analog resistant to enzymatic degradation and
possibly increase its antiproliferative activity. To prepare xylo-
adenylyl-(2'→5')-xyloadenylyl-(2'→5')-xyloadenosine 5 (17) we selected the phosphotriester approach in liquid phase. Our strategy
required the large-scale synthesis of two nucleosidic synthons 6
and 7, of which 7 could be obtained from 6 in two additional steps
(89) (Fig. 2).

Although synthon 6 can be prepared from 9-β-D-xylofuranosyl-
adenine, this later nucleoside is not commercially available.
Therefore, it was necessary to synthesize a protected derivative
of this nucleoside according to our previous strategy, i.e., use

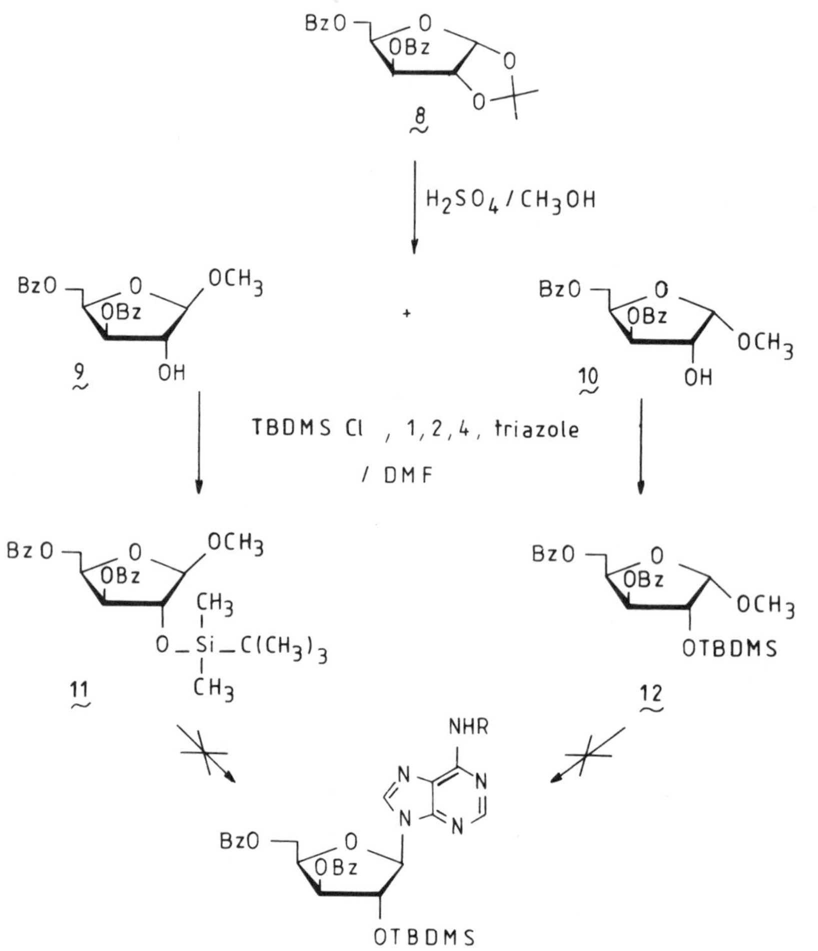

_ Scheme III _

of a properly protected xyloside prior to glycosylation. This
approach necessitates the following conditions (90) : 1) the dissymmetrically substituted xylofuranose must be easily synthesized
in good yield, 2) the protecting groups must be stable enough to
withstand glycosylation and subsequent reaction at the nucleotide
stage, and 3) the method of glycosylation should lead exclusively

TABLE III

AGLYCONE	CATALYST	SOLVENT (CONDITIONS)
Adenine	$SnCl_4$	Acetonitrile; reflux or room temperature (93)
	$C_4F_9SO_3K$	Hexamethyldisilazane, trimethyl-chlorosilane, acetonitrile; reflux (94-96)
N-Benzoyladenine	$p-CH_3-C_6H_4-SO_3H$	Fusion; 20 mn; $172^{o}C$; 20 mm Hg (97)
Bistrimethylsilyl N-benzoyladenine	$SnCl_4$	1,2-dichloroethane; reflux or room temperature (98-99)
	CF_3SO_2TMS	Dichloromethane; reflux or room temperature (95)
Chloromercuri N-benzoyladenine	$TiCl_4$	Dichloromethane; celite; reflux (100-101)

to the desired 9-β-N-isomer.

Regarding the first point, the approach we used to prepare the dissymmetrical ribofuranoses (13) described in part I was not applicable, as such analogs in the xylo series, e.g., 1,3,5-tri-O-acyl-D-xylofuranose, are extremely difficult to prepare (91). In addition to the stability of the protecting groups, mentioned in the second point, they must be capable of being selectively removed at specific points in our synthetic sequence without causing any problems. The TBDMS group met these requirements (92) and we used it to synthesize the 2-O-TBDMS xylosides 11 and 12. These anomers could be conveniently and rapidly prepared from 1,2-di-O-isopropylidene-3,5-di-O-benzoyl-α-D-xylofuranose (8, Scheme III). Unfortunately, condensation of either 11 or 12 with adenine (or a suitably protected derivative), using various conditions and catalysts (Table III), did not provide the desired nucleoside.

Therefore, we changed our initial approach and decided to

introduce the 2'-O-TBDMS group after glycosylation had been performed. We chose the xylofuranose 13 as our starting material.
This sugar derivative contains two desirable features ; the acyl
groups on positions 1 and 2. The 1-O-acetyl group should facilitate glycosylation and the 2-O-acetyl, due to participation as an
acetoxonium ion, should influence the formation of the β-anomer.
In addition, we can also differentiate between the 2' and 3' positions of the anticipated nucleoside as the 2'-O-acetyl group is
preferentially hydrolyzed over the benzoyl functions. Starting
with 8, we were able to prepare 13 in one step and in quantitative
yield.

Condensation of 13 (α anomer) with adenine using 2 equivalents of SnCl$_4$ as catalyst (93) furnished the β-xylofuranosylade-
nine derivative 14 in 70 % yield. As expected, hydrazinolysis
(102) of this nucleoside 14 gave 15 in near-quantitative yield. At
this point, we introduced the TBDMS group on position 2 by the
conventional procedure (103). The resulting derivative 16 was then

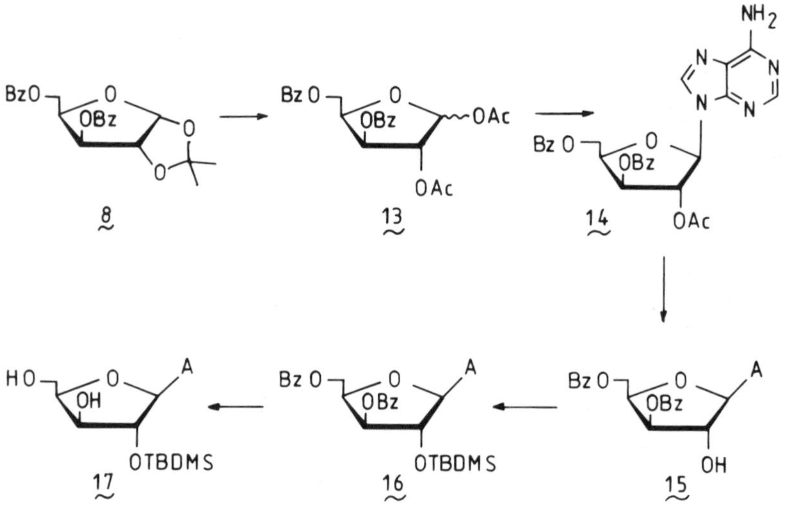

Scheme IV

selectively debenzoylated with sodium methoxide to furnish the
precursor 17 of the desired synthons 6 and 7. It is noteworthy
that the trans orientation of the secondary hydroxyl groups on
compound 15 and 17 prevents any migration between the 2' and 3'
positions (Scheme IV).

Next, the primary hydroxyl of 17 was protected with a monomethoxytrityl group and the resulting derivative 18, was treated
with benzoic anhydride / pyridine in the presence of 4-dimethyla-
mino pyridine (104) to provide the completely protected nucleoside 19.

Subsequent removal of the TBDMS group by tetrabutylammonium
fluoride gave the first synthon 6, which was then successively
benzoylated and detritylated to afford the second synthon 7. The
nucleoside 6 also served as the precursor of 21 and was conveniently converted to this analog when reacted with an excess of o-
chlorophenyl phosphorodi-(1,2,4-triazolide (105) (Scheme V).

For oligonucleotide condensation, we selected 1-mesitylene-
sulfonyl-3-nitro-1,2,4-triazole (MSNT) as the activating agent because of its reported stability and effectiveness (106-107).
Reacting 7 and 21 in the presence of MSNT gave 22 which was directly
detritylated to furnish 23. Once chromatographically pure, 23 was
in turn, reacted with 21 to provide the protected trimer 24 in
good yield (Scheme VI).

Finally, the desired core 5 was obtained by treating the trimer 24 successively and in order with : i) tetramethylguanidinium-
syn-4-nitrobenzaldoximate (106-107) to remove o-chlorophenyl
groups ; ii) aqueous ammonia to hydrolyse the benzoyls ; and iii)
80 % acetic acid to remove the monomethoxytrityl group. The chromatographic purification of 5 was accomplished on a DEAE Sephadex
A-25 column using a linear gradient of ammonium hydrogen carbonate.
The purity of 5 was verified by TLC, HPLC, enzymatic digestions
and high resolution (500 MH$_z$) NMR spectroscopy.

The biological data concerning this core 5 are quite interesting (108). The stability of 5 against degradation by cellular

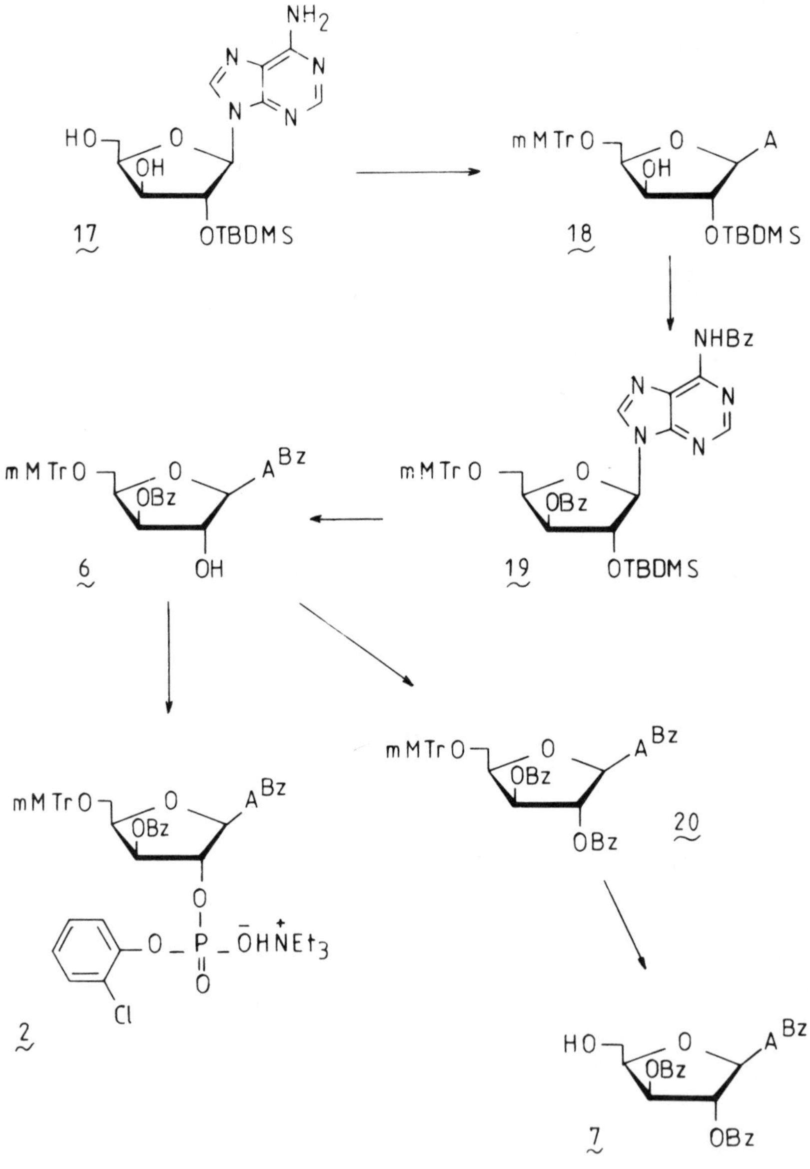
NH2
HO
O
OH
17
OTBDMS
mMTrO
O
A
OH
18
OTBDMS
NHBz
mMTrO
O
ABz
OBz
6
OH
mMTrO
O
OBz
19
OTBDMS
mMTrO
O
ABz
OBz
20
OBz
mMTrO
O
ABz
OBz
O
O — P — OHNEt3
O
Cl
2
HO
O
ABz
OBz
7
OBz

Scheme V

phosphodiesterase was approximately 120 times greater than that of the natural 2-5 A core, as determined by incubation in L cell-free extracts followed by HPLC analysis (Fig. 3).

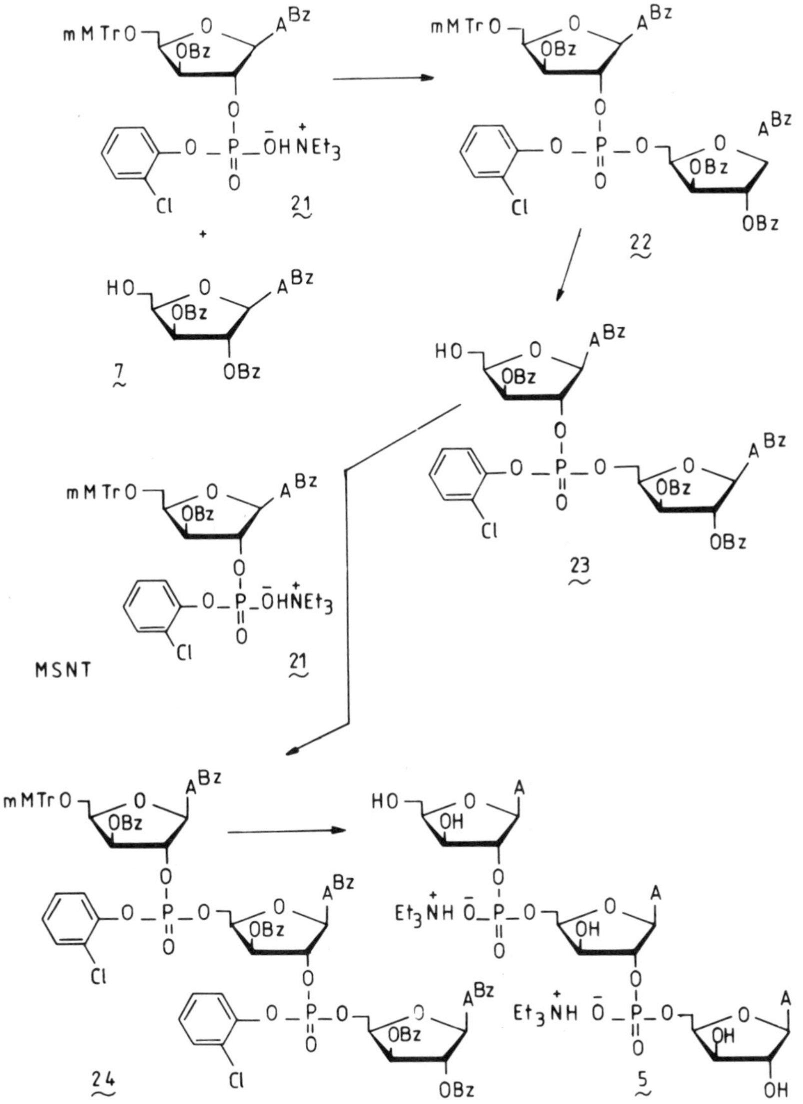

Scheme VI

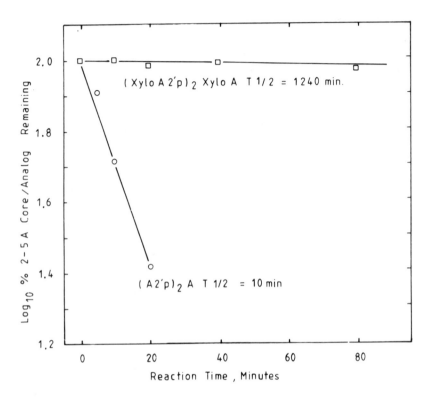

FIGURE 3.   Stability of (xyloA2'p)$_2$xyloA and (A2'p)$_2$A in
murine L cell-free extract. Xylo core analog or natural 2-5 A core
(P.L. Biochemicals) was incubated at 10^{-4}M at 30°C with 1/3 volume
of L-cell S-10 extract (55) containing 20 mM HEPES (pH 7.4),
120 mM KCl, 5 mM Mg(OAc)$_2$, and 1mM DTT. Reactions were terminated
after 5-320 min by heating at 95°C for 5 min, samples were cooled
to 4°C, centrifuged 5 min at 10,000 **×** g, and the supernatants were
analyzed by HPLC on a 30 **×** 0.4 cm column packed with octadecylsi-
lane. C 18 bonded to 10 μ silica (Phase-Separations Limited, U.K.).
The column was eluted at 1.5 ml/min with mobile phase consisting
of 200 mM ammonium phosphate (pH 7.0), with a linear gradient of
0-20% methanol in 20 min. Absorbance at 260 nm was monitored, and
peak area was determined by integration. Degradation was calculated as the percent of zero-time peak remaining, and half-life was
calculated from a first order rate equation.

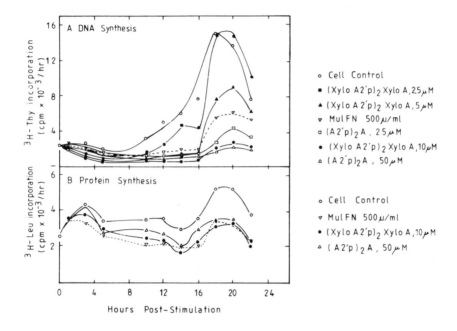

FIGURE 4.   Cell-growth inhibition by (xyloA2'p)₂xyloA and (A2'p)₂A in synchronized, serum-stimulated 3T3 fibroblasts. Swiss albino 3T3 cells, grown in DME containing 10% calf serum, were plated in 96 well microtiter plates (1 ✗ 10⁴ cells/well), incubated until cells were 90-95% confluent (28 hr), and then arrested in G₀ by incubation for 24 hr with conditioned medium (fresh DME with 10% of medium from confluent cultures). 2-5A core, xylo core, or mouse fibroblast interferon (2.2 ✗ 10⁷ U/mg, prepared by Lee Biomolecular, San Diego, California, and obtained from M. Krim at the Memorial Sloan-Kettering Interferon Labs, New York) were added to quadruplet wells together with fresh medium containing 10% serum to stimulate the entry into the cell cycle. (A) DNA or (B) protein synthesis was determined at the indicated times by washing, pulse-labeling for 1 hr respectively with (³H) Thy or (³H) Leu (0.5 μCi/0.2 ml/well), and determining TCA-precipitable radioactivity (14). O, control; □ , 25 μM and Δ, 50 μM (A2'p)₂A; ∎, 2.5 μM, and ▲, 5 μM, and ●, 10 μM (xyloA2'p)₂xyloA; ∇, MuIFN-β, 500 U/ml.

TABLE IV. *Effect of 2-5A Core, Xylo 2-5A Core, or Interferon on Cellular DNA synthesis*[a]

A. Swiss 3T3 Cells.

	16 hr		18 hr		20 hr		Total	
	cpm[b]	% inhib	cpm	% inhib	cpm	% inhib	cpm	% inhib
Control	7,548	0	14,813	0	13,475	0	35,836	0
2-5A core :								
50 μM	1,084	86	1,538	90	2,097	84	4,719	87
25 μM	1,426	81	2,645	82	4,260	68	8,331	77
Xylo 2-5A Core :								
10 μM	715	91	2,074	86	2,704	80	5,493	85
5 μM	1,673	78	7,510	49	8,833	34	18,016	50
2.5 μM	4,378	42	14,788	0	14,603	0	33,769	6
Mouse Interferon :								
500 U/ml	1,890	75	5,491	63	6,027	55	13,388	63

B. NIH-3T3 Clone 1 Cells[c]

	18 hr		20 hr		22 hr		Total	
	cpm[b]	% inhib	cpm	% inhib	cpm	% inhib	cpm	% inhib
Control	29,432	0	33,996	0	27,051	0	90,479	0
2-5A Core :								
100 μM	4,090	86	5,246	85	5,370	80	14,706	84
50 μM	4,854	84	5,335	84	8,498	67	18,687	79
25 μM	15,728	47	23,675	30	36,897	0	76,300	18
Xylo 2-5A Core :								
10 μM	266	99	394	99	236	99	896	99
5 μM	1,092	96	850	98	1,516	94	3,458	96
2.5 μM	2,425	92	3,591	89	6,188	77	12,204	87
1 μM	14,631	50	15,941	53	19,537	28	50,109	45
Mouse Interferon :								
500 U/ml	28,908	0	30,425	11	31,000	0	90,333	0

[a]Experimental conditions were as described in Fig. 4. DNA synthesis in control cultures peaked at 18 or 20 hr post-stimulation in Swiss or NIH-3T3 cells, respectively.

[b](^{3}H)Thy incorporation/60 min into acid-precipitable material was measured. All results are the average of four replicates.

[c]NIH-3T3, clone 1, cells were obtained from Dr Robert Friedman, USUHS, Bethesda, MD.

In intact cells, the antiproliferative activity of 5 was determined according to the general method of Kimchi et al., (29) in synchronized serum stimulated Swiss 3T3 normal fibroblasts and NIH-3T3 clone 1 cells. The latter cells have many unique characteristics, e.g., they are defective in the 2-5A dependent endonuclease and resistant to the antiproliferative action of interferons and antiviral action of interferons on lytic viruses (57,109). The xylo core 5 was shown to be approximately five times more active than the natural 2-5A core in inhibiting DNA synthesis in the Swiss cells (Fig. 4 and Table IV) and approximately twenty-forty times more active than natural core in the NIH-3T3 cells (Table IV). The inhibition of protein synthesis was much less than was the inhibition of DNA synthesis (Fig. 4). The antiviral properties of 5 have also been determined (110). Xylo 2-5A core was effective against several DNA but not RNA viruses ; thus in Vero cells, the ED_{50} of xylo 2-5A core against herpes simplex viruses 1 and 2 was found to be comparable to that obtained with Acyclovir. Xyloadenosine, which is a degradation product of xylo 2-5A core, was active only with at least ten-fold higher concentrations.

We do not know the precise mechanism by which this $(xyloA2'p)_2$ xyloA core 5 results in inhibition of DNA synthesis; neither the xylo 2-5A core nor the parent 2-5A core result in activation of the 2-5A dependent endonuclease as originally speculated (111). However, the fact that without artificial permeabilization of the cells this compound is able to effectively inhibit growth in cells that are resistant to the antiproliferative activity of interferons has intriguing implications for anticancer and antiviral applications.

IV. CONCLUSION

Oligonucleotide synthesis necessitates - in somes cases - preparation of appropriate nucleosidic synthons and subsequent selective protection of various functional groups on the sugar moiety and aglycone.

We have proposed a general synthetic approach for reaching such nucleosidic derivatives by introducing a 2',3'-differentiation on the starting sugar.

We have presented, in conjunction with other data, our strategy for the synthesis of the $(xyloA2'p)_2xyloA$.

This compound has been evaluated for its antimitogenic activity and has been shown to be a much more potent as inhibitor of cell growth than the natural 2,5A core.

This is also the first 2-5A core analog that has been shown *in vitro* to effectively inhibit replication of several DNA viruses.

ACKNOWLEDGMENTS

The authors thank Julien P.H. Verheyden (Syntex) for helpful discussions.

REFERENCES

1. Reese, C.B., *Tetrahedron 34*, 3143 (1978).
2. Amarnath, V., and Broom, A.D., *Chem. Rev. 77*, 183 (1977).
3. Ikehara, M., Ohtsuka, E., and Markhas, A.F., *Adv. Carbohydr. Chem. Biochem. 36*, 135 (1979).
4. Van Boom, J.H., and Burgers, P.M.J., *Tetrahedron Lett.* 4875, (1976).
5. Ogilvie, K.K., Beaucage, S.L. Schifman, A.L., Theriault, N.Y., and Sadana, K.L., *Can. J. Chem. 56*, 2768 (1978).
6. Warkiewicz, W.T., and Wierviorowski, M., *Nucleic Acid Res. 2*, 951 (1975).
7. Hall, R.H., *The Modified Nucleosides in Nucleic Acids*, Columbia University Press, New York, 1971.
8. Kohler, W., Schlosser, W., Charubala, G., and Pleiderer, W., in *Chemistry and Biology of Nucleosides and Nucleotides*, (R.E. Harmon, R.K. Robins and L.B. Townsend, eds.), Academic Press, New York, 1978, pp. 347-358.
9. Einck, J.J., Pettit, G.R., Brown, P., and Yamauchi, K., *J. Carbohydr. Nucleosides, Nucleotides 7*, 1 (1980).
10. Ekborg, G., and Garegg, P.J., *J. Carbohydr. Nucleosides, Nucleotides 7*, 57 (1980).
11. Ness, R.K., and Fletcher, H.G., Jr., *J. Am. Chem. Soc. 78*, 4710 (1954).
12. Dumont, F., Barascut, J.L., Chavis, C., and Imbach, J.L., in *Lectures in Heterocyclic Chemistry*, 5, S-27 (1980).
13. Chavis, C., Dumont, F., and Imbach, J.L., *J. Carbohydr. Nucleosides, Nucleotides 5*, 133 (1978).

14. Ogilvie, K.K., and Entwistte, D.W., *Carbohydr. Res. 89,* 203, (1981).
15. Reese, C.B., and Trentham, D.R., *Tetrahedron Lett.* 2467 (1965)
16. Maki, Y., Kihuchi, K., Sugiyama, H., and Seto, S., *Tetrahedron Lett.* 3295 (1975).
17. Gosselin, G., and Imbach, J.L., *Tetrahedron Lett. 22,* 4699 (1981).
18. Stewart II W.E., *The Interferon System,* 2nd edition. Springer Verlag, New York, 1981.
19. Samuel, C.E., Farris, D.F., and Eppstein, D.A., *Virology 83,* 56 (1977).
20. Williams, B.R.G., and Kerr, I.M., *Nature 276,* 88 (1978).
21. Williams, B.R.G., Golgher, R.R., and Kerr, I.M., *FEBS Lett. 105,* 47 (1979).
22. Williams, B.R.G., Golgher, R.R., Brown, R.E., Gilbert, C.S. and Kerr, I.M., *Nature 282,* 582 (1979).
23. Hovanessian, A.G., Wood, J., Meurs, E., and Montagnier, L., *Proc. Natl. Acad. Sci. USA 76,* 3261 (1979).
24. Baglioni, C., Maroney, P.A., and West, D.K., *Biochemistry 18,* 1765 (1979).
25. Nilsen, T.W., Wood, D.L., and Baglioni, C., *Nature 286,* 176 (1980).
26. Hovanessian, A.G., and Wood, J.N., *Virology 101,* 81 (1980).
27. Kimchi, A., Shure, H., and Revel, M., *Nature 282,* 849 (1979).
28. Kimchi, A., Shure, H., and Revel, M., *Eur. J. Biochem. 114,* 5 (1981).
29. Kimchi, A., Shure, H., Lapidot, Y., Rapoport, S., Panet, A., and Revel, M., *FEBS Lett. 134,* 212 (1981).
30. Krishnan, I., and Baglioni, C., *Virology 111,* 666 (1981).
31. Leanderson, T., Nordfelth, R., and Lundgren, E., *Biochem. Biophys. Res. Commun. 107,* 511 (1982).
32. Williams, B.R.G., Kerr, I.M., Gilbert, C.S., White, C.N., and Ball, L.A., *Eur. J. Biochem. 92,* 455 (1978).
33. Clemens, M.J., and Williams, B.R.G., *Cell 13,* 565 (1978).
34. Eppstein, D.A., and Samuel, C.E., *Virology 89,* 240 (1978).
35. Baglioni, C., Minsk, M.A., and Maroney, P.A., *Nature 273,* 684 (1978).
36. Ratner, L., Wiegand, R.C., Farrel, P.J., Sen, G.C., Cabrer, B., and Lengyel, P., *Biochem. Biophys. Res. Commun. 81,* 947 (1978).
37. Schmidt, A., Zilberstein, A., Shulman, L., Federman, P., Berissi, H., and Revel, M., *FEBS Lett. 95,* 257 (1978).
38. Ball, L.A., and White, C.N., *Virology 93,* 348 (1979).
39. Slattery, E., Ghosh, N., Samantha, H., and Lengyel, P., *Proc. Natl. Acad. Sci. USA 76,* 4778 (1979).
40. Vaquero, M., and Clemens, M.J., *Eur. J. Biochem. 98,* 245 (1979).
41. Nilsen, T.W., Weissman, S.G., and Baglioni, C., *Biochemistry 19,* 5574 (1980).

42. Johnston, M.I., Friedman, R.M., and Torrence, P.F., *Biochemistry 19*, 5580 (1980).
43. Minks, M.A., Benvin, S., and Baglioni, C., *J. Biol. Chem. 255*, 5031 (1980).
44. Shulman, L., and Revel, M., *Nature 287*, 98 (1980).
45. Justesen, J., Ferbus, D., and Thang, M.N., *Proc. Natl. Acad. Sci. USA 77*, 4618 (1980).
46. Miyamoto, N.G., and Samuel, C.E., *Biochem. Biophys. Res. Commun. 101*, 680 (1981).
47. Nilsen, T.W., Maroney, P.A., and Baglioni, C., *J. Biol. Chem. 256*, 7806 (1981).
48. Yang, K., Samanta, H., Dougherty, J., Jayaram, B., Broeze, R., and Lengyel, P., *J. Biol. Chem. 256*, 9324 (1981).
49. Floyd-Smith, G., Slattery, E., and Lengyel, P., *Science 212*, 1030 (1981).
50. Baglioni, C., Minks, M.A., and De Clercq, E., *Nucleic Acids Res. 9*, 4939 (1981).
51. Besançon, F., Bourgeade, M.F., Justesen, J., Ferbus, D., and Thang, M.N., *Biochem. Biophys. Res. Commun. 103*, 16 (1981).
52. Koliais, S.I., and Kortsaris, A., *J. Gen. Virol. 58*, 191 (1982).
53. Lab, M., Thang, M.N., Soteriadou, K., Koehren, F., and Justesen, J., *Biochem. Biophys. Res. Commun. 105*, 412 (1982).
54. Ball, L.A., in *The Enzymes*, vol. XV, Academic Press, 1982, pp. 282-313.
55. Eppstein, D.A., Peterson, T.C., and Samuel, C.E., *Virology 98*, 9 (1979).
56. Schmidt, A., Chernajovsky, Y., Shulman, L., Federman, P., Berissi, H., and Revel, M., *Proc. Natl. Acad. Sci. USA 76*, 4788 (1979).
57. Panet, A., Czarniecki, C.W., Falk, H., and Friedman, R.M., *Virology 114*, 567 (1981).
58. Drocourt, J.L., Dieffenbach, C.W., and Ts'O, P.O.P., *Nucleic Acids Res. 10*, 2163 (1982).
59. Ogilvie, K.K., and Theriault, N.Y., *Tetrahedron Lett.*, 2111 (1979).
60. Ikehara, M., Oshie, K., and Ohtsuka, E., *Tetrahedron Lett.*, 3677 (1979).
61. Sawai, H., Shibata, T., and Ohno, M., *Tetrahedron Lett.*, 4573 (1979).
62. Engels, J., and Krahmer, U., *Angew. Chem. Int. Ed. 18*, 942 (1979).
63. Den Hartog, J.A.J., Doornbos, J., Crea, R., and Van Boom, J.H., *Recl. Trav. Chim. Pays Bas 98*, 469 (1979).
64. Martin, E.M., Birdsall, N.J.M., Brown, R.E., and Kerr, I.M., *Eur. J. Biochem. 95*, 295 (1979).
65. Markham, A.F., Porter, R.A., Gait, M.J., Sheppard, R.C., and Kerr, I.M., *Nucleic Acids Res. 6*, 2569 (1979).
66. Jones, S.S., and Reese, C.B., *J. Amer. Chem. Soc. 101*, 7399 (1979).

67. Charubala, R., and Pfleiderer, W., *Tetrahedron Lett.*, 1933 (1980).
68. Chattopadhyaya, J.B., *Tetrahedron Lett.*, 4113 (1980).
69. Den Hartog, J.A.J., Wijnands, R.A., and Van Boom, J.H., *Nucleic Acids Res., Symposium Series n° 7*, 157 (1980).
70. Giveli, C., Kwiatkowski, M., Oberg, B., and Chattopadhyaya, J.B., *Tetrahedron Lett.*, 1741 (1981).
71. Charubala, R., Uhlmann, E., and Pfleiderer, W., *Liebigs Ann. Chem.*, 2392 (1981).
72. Sawai, H., Shibata, T., and Ohno, M., *Tetrahedron 37*, 481 (1981).
73. Ikehara, M., Oshie, K., Hasegawa, A., and Ohtsuka, E., *Nucleic Acids Res. 9*, 2003 (1981).
74. Den Hartog, J.A.J., Wijnands, R.A., and Van Boom, J.H., *J. Org. Chem. 46*, 2242 (1981).
75. Imai, J., and Torrence, P.F., *J. Org. Chem. 46*, 4015 (1981) ; in *Methods in Enzymology*, Vol. 79, Academic Press, 1981, pp. 233-244.
76. Ohtsuka, E., Yamane, A., and Ikehara, M., *Chem. Pharm. Bull. 30*, 376 (1982).
77. Karpeisky, M.Y., Beigelman, L.N., Mikhailov, S.N., Padyukova, N.S., and Smrt, J., *Collect. Czech. Chem. Commun. 47*, 156 (1982).
78. Knight, M., Cayley, P.J., Silverman, R.H., Wreschner, D.H., Gilbert, C.S., Brown, R.E., and Kerr, I.M., *Nature 288*, 189 (1980).
79. Kerr, I.M., and Brown, R.E., *Proc. Natl. Acad. Sci. USA 75*, 256 (1978).
80. Doetsch, P.W., Suhadolnik, R.J., Sawada, Y., Mosca, J.D., Flick, M.B., Reichenbach, N.L., Dang, A.Q., Wu, J.M., Charubala, R., Pfleiderer, W., and Henderson, E.E., *Proc. Natl. Acad. Sci. USA 78*, 6699 (1981).
81. Gazitt, Y., *Cancer Res. 41*, 2959 (1981).
82. Charubala, R., and Pfleiderer, W., *Tetrahedron Lett.*, 4077 (1980).
83. Engels, J., *Tetrahedron Lett.*, 4339 (1980).
84. Doetsch, P.W., Wu, J.M., Sawada, Y., and Suhadolnik, R.J., *Nature 291*, 355 (1981).
85. Kwiatkowski, M., Gioeli, C., Oberg, B., and Chattopadhyaya, J.B., *Chemica Scripta 18*, 95 (1981).
86. Kwiatkowski, M., Gioeli, C., Chattopadhyaya, J.B., Oberg, B., and Drake, A.F., *Chemica Scripta 19*, 49 (1982).
87. Jager, A., and Engel, J., *Nucleic Acids Res., Symposium Series n° 9*, 149 (1981).
88. Baglioni, C., D'Alessandro, S.B., Nilsen, T.W., Den Hartog, J.A.J., Crea, R., and Van Boom, J.H., *J. Biol. Chem. 256*, 3253 (1981).
89. Gosselin, G., and Imbach, J.L., *J. Heterocyclic Chem. 19*, 597 (1982).

90. Dumont, F., Wightman, R.H., Ziegler, J.C., Chavis, C., and Imbach, J.L., *Tetrahedron Lett.* 3291 (1979).
91. Stevens, J.D., and Fletcher, H.G., *J. Org. Chem. 33*, 1799 (1968).
92. Chavis, C., Dumont, F., Wightman, R.H., Ziegler, J.C., and Imbach, J.L., *J. Org. Chem. 47*, 202 (1982).
93. Saneyoshi, M., and Satoh, E., *Chem. Pharm. Bull. 27*, 2518 (1979).
94. Vorbruggen, H., and Bennua, B., *Tetrahedron Lett.*, 1339 (1978).
95. Vorbruggen, H., Krolikiewicz, K., and Bennua, B., *Chem. Ber. 114*, 1234 (1981).
96. Vorbruggen, H., and Bennua, B., *Chem. Ber. 114*, 1279 (1981).
97. Sako, T., Shimadate, T., and Ishido, Y., *Nippon Kagaku Zasshi 81*, 1440 (1962).
98. Niedballa, U., and Vorbruggen, H., *J. Org. Chem. 39*, 3654 (1974).
99. Lichtenthaler, F.W., Voss, P., and Heerd, A., *Tetrahedron Lett.*, 2141 (1974).
100. Tong, G.L., Lee, W.W., and Goodman, L., *J. Org. Chem. 32*, 1984 (1967).
101. Bouchu, D., Abou-Assali, M., Grouiller, A., Carret, G., and Pacheco, H., *Eur. J. Med. Chem. 16*, 43 (1981).
102. Ishido, Y., Sakairi, N., Okazaki, K., and Nakazaki, N., *J. Chem. Soc. Perkins Trans I*, 563 (1980).
103. Ogilvie, K.K., Schifman, A.L., and Penney, C.L., *Can. J. Chem. 57*, 2230 (1979).
104. De Rooij, J.F.M., Wille-Hazeleger, G., Van Deursen, P.H., Serdijn, J., and Van Boom, J.H., *Recl. Trav. Chim. Pays Bas 98*, 537 (1979).
105. Chattopadhyaya, J.B., and Reese, C.B., *Tetrahedron Lett.* 5059 (1979) ; *Nucleic Acids Res. 8*, 2039 (1980).
106. Reese, C.B., Titmus, R.C., and Yan, L., *Tetrahedron Lett.* 2727 (1978).
107. Jones, S.S., Rayner, B., Reese, C.B., Ubasawa, A., and Ubasawa, M., *Tetrahedron 36*, 3075 (1980).
108. Eppstein, D.A., Marsh, Y.V., Schryver, B.B., Larsen, M.A., Barnett, J.W., Verheyden, P.H., and Prisbe, E.J., *J. Biol. Chem.*, in press (1982).
109. Eppstein, D.A., Czarniecki, C.W., Jacobsen, H., Friedman, R.M., and Panet, A., *Eur. J. Biochem. 118*, 9 (1981).
110. Eppstein, D.A., Barnett, J.W., Marsh, Y.V., Gosselin, G., and Imbach, J.L., manuscript in preparation.
111. Eppstein, D.A., Schryver, B.B., Marsh, Y.V., and Larsen, M.A., manuscript in preparation.

SELECTIVE MODIFICATION AND DEOXYGENATION

AT C2' OF NUCLEOSIDES[1]

Morris J. Robins, Fritz Hansske, John S. Wilson,

S.D. Hawrelak, and Danuta Madej

Department of Chemistry

The University of Alberta

Edmonton, Alberta, Canada

I. INTRODUCTION

We have been interested in the development of generally
applicable methods for selective deoxygenation and other
modifications of the sugar moiety of nucleosides for several
years (1-12). These transformations allow conservation of
structural and stereochemical features of the naturally
occurring precursor nucleosides. They also may be designed to
mimic biosynthetic routes to other natural products. A relatively small number of successful approaches at C2' that are
independent of the structure and participation of the base
moiety have been reported.

Pyrimidine O2→C2' cyclonucleosides (2,2'-anhydro arabinonucleosides) have been utilized extensively for ribo to

[1]This work was supported by the Natural Sciences and
Engineering Research Council of Canada, the National
Cancer Institute of Canada, and The University of Alberta.

arabinonucleoside conversions, for efficient preparations of
2'-substituted-2'-deoxyribonucleosides, and as intermediates
for the preparation and hydrogenolysis of 2'-halo-2'-deoxy
compounds to give 2'-deoxynucleosides (13-15). However, the
use of analogous sequences with purine O8→C2' cyclonucleosides
(8,2'-anhydro arabinonucleosides) results in retention of an
"unnatural" substituent at C8, whose removal can be troublesome
(16-18). In addition, some nucleoside analogues do not have
appropriate base structures to accommodate the cyclonucleoside
approach. A recent review by Moffatt documents the considerable effort that has been directed toward modification and
deoxygenation of the sugar moiety of nucleosides and the
limited number of successful selective conversions at C2' (19).

A major continuing problem had been the differentiation of
the two secondary cis hydroxyl functions to allow specific
protection of O3' and O5'. This was solved recently by the
introduction of 1,3-dichloro-1,1,3,3-tetraisopropyldisiloxane
by Markiewicz and Wiewiórowski (20,21). Treatment of nucleosides with this hindered bifunctional silylating reagent in
pyridine gives the 3',5'-$\underline{O}$-(1,1,3,3-tetraisopropyldisilox-1,3-
diyl) (3',5'-$\underline{O}$-TPDS) derivatives in high yields (12,21).

The second major obstacle in the 2'-modification of nucleosides results from inherent difficulties with reactivity at
C2'. The electronegative C1' anomeric center (an $\underline{N}$,O-acetal
carbon) bonded to C2' effectively prevents ionization of a 2'-
leaving group to produce a cationic S_N1 intermediate. Related
S_N2 type displacements in which bond cleavage to the leaving
group is advanced relative to bond formation with the nucleo-
phile are faced with analogously severe energy barriers. This
trend is apparent with intramolecular displacements of chloride
at C2' $\underline{vs}$. C3'. Conversion of 9-(2-chloro-2-deoxy-β-D-arabino-
furanosyl)adenine to 2',3'-anhydroadenosine requires more
vigorous conditions and/or extended times relative to the

analogous reaction of 9-(3-chloro-3-deoxy-β-D-xylofurano-
syl)adenine to give the same ribo-epoxide product (4,22).
These nucleophilic ring closures both involve intramolecular
attack of the trans-oriented neighboring hydroxyl oxygen from
the sterically unhindered α-face of the sugar ring.

A major steric barrier to attack of "external" nucleophiles
on C2′ at the β-face is exerted by the heterocyclic base at C1′
of β-ribonucleosides. In addition, formation of the resulting
β-arabinonucleoside product results in introduction of parallel
dipoles at C1′ and C2′. Combinations of steric and electronic
factors contribute to the different ratios of C2′ (arabino) and
C3′ (xylo) halo-substituted deoxynucleosides formed by nucleophilic attack of halides on ribonucleoside 2′,3′-acetoxonium
(dioxolenium) ion intermediates (1,2,4-7,22,23). These ratios
of C2′:C3′ isomers vary from ~0:100 with 4-amino-7-(β-D-
ribofuranosyl)pyrrolo[2,3-d]pyrimidine (tubercidin, 7-deaza-
adenosine) (1,6,23), to ~1:10 with adenosine (4,5,22), to
~35:65 with 7-amino-3-(β-D-ribofuranosyl)pyrazolo[4,3-d]-
pyrimidine (formycin) (1,23). Tubercidin (1) (see Figure 1) is
known from X-ray crystallographic analysis to have a very short
glycosyl bond (C1′—N7: 1.438 Å) (24), which would enhance the
steric hindrance to nucleophilic attack by halide on C2′ at the
β-face. Adenosine (2) has a longer glycosyl bond (C1′—N9:
1.466 Å) (25) than (1), but a similar N,O-acetal electronic

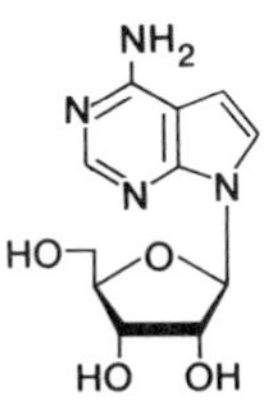

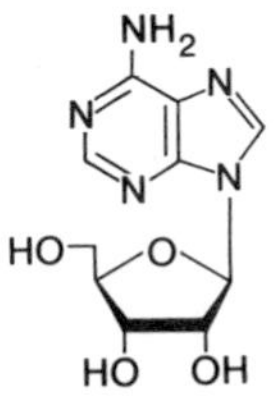

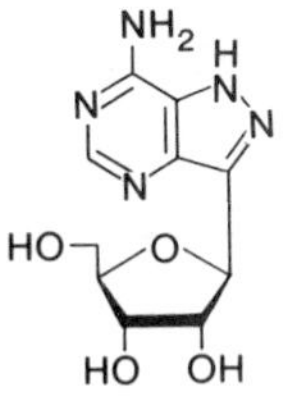

FIGURE 1

demand at C1'. Formycin (3) has the longest sugar to
heterocycle bond (C1'—C3: 1.501 Å) (26) plus the favorable
electronic factor that C1' is at the ether level of oxidation
in this C-nucleoside.

Limited success had been realized employing displacement of
the 2'-O-chlorosulfite group (generated in situ) with chloride
(10). Nucleophilic displacements of 2'-triflate proceeded efficiently, but multistep routes were required for preparation
of their protected precursors (18,27,28). Attempted reduction
of 2'-O-(tetra-N-methylphosphorodiamidate) derivatives with
dissolving metals in liquid amines (29,30) resulted in rapid
elimination of the heterocyclic base.[2] Other observations of
liberation of the weakly basic heterocycle from C1' upon generation of carbanion character at C2' have been reported (31,32).

Free radical mediated reductions of 2'-halo-2'-deoxy-
nucleosides using tri-n-butyltin hydride with initiation by
α,α'-azobisisobutyronitrile (AIBN) had been demonstrated to
give 2'-deoxynucleosides cleanly without complications
(4,6,10,15). Enzymatic reduction of ribonucleoside diphosphates by the ribonucleotide reductase from Escherichia coli
was known to employ a long-lived free radical in the B2-subunit
(33). We therefore concentrated our efforts on homolytic
approaches to direct deoxygenation.

Barton and co-workers had deoxygenated secondary alcohols
by treatment of thiocarbonyl derivatives with tri-n-butyltin
hydride (34,35). They described reduction of a protected
adenosine cyclic 2',3'-thionocarbonate to give a mixture of 2'-
and 3'-deoxyadenosine (35). Use of the non-initiated Barton
reaction conditions with thionobenzoate esters had given benzyl

---

[2]M.J. Robins and P. Sporns, unpublished results.

ether (36) and starting alcohol by-products (34,36). Analogous
reduction of methyl dithiocarbonate (xanthate) esters precluded
benzyl ether formation, but reversion to the starting alcohol
remained a significant side-reaction (34,36). Barton had reported that conversion of tri-n-butyltin hydride to hexa-n-
butyldistannane occurred during sluggish reductions of certain
thiocarbonylimidazolide derivatives of alcohols, presumably
catalyzed by imidazole released during the reduction (34).
Therefore, we sought a thiocarbonyl reagent that would avoid
this potential problem. We have employed phenyl chlorothionocarbonate (12) as a reagent that functions as a reasonably
active acylating agent and gives phenoxythiocarbonyl esters
that are stable and easily identified spectrally. Reduction of
these phenyl thionocarbonates proceeds cleanly and in high
yields (12). Others have reported deoxygenations of nucleoside
analogues using Barton procedures (36-39).

II. SELECTIVE 2'-DEOXYGENATION OF NUCLEOSIDES

Treatment of naturally occurring ribonucleosides (4) (see
Figure 2) with 1,3-dichloro-1,1,3,3-tetraisopropyldisiloxane in
pyridine generally resulted in smooth and high-yield formation
of 3',5'-O-(1,1,3,3-tetraisopropyldisiliox-1,3-diyl)-nucleo-
sides (3',5'-O-TPDS-nucleosides) (5). The low solubility of
guanosine in pyridine necessitated the use of N,N-dimethyl-
formamide (DMF) as co-solvent. By-product formation occurred
in this case and 3',5'-O-TPDS-guanosine was obtained in only
70% yield. Migration of the TPDS group also was observed by
van Boom and coworkers with guanosine in DMF with hydrogen
chloride present (40). Markiewicz and Wiewiórowski have
studied migration equilibria of several TPDS derivatives in DMF
containing hydrogen chloride and other acids (41).

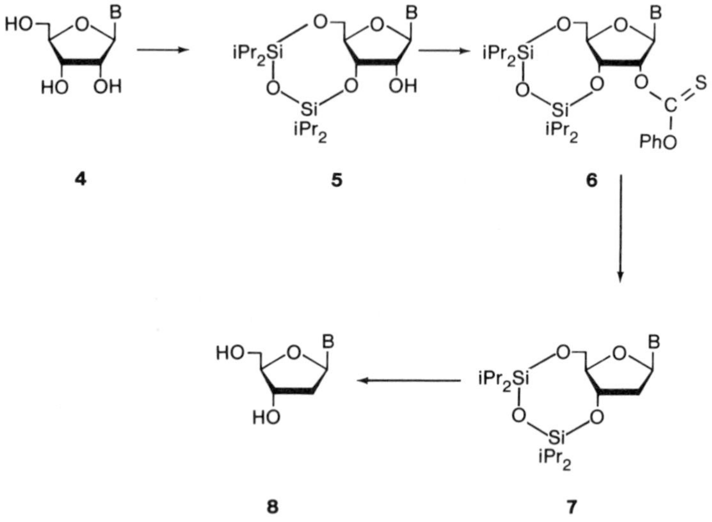

FIGURE 2

Treatment of 3',5'-O-TPDS-nucleosides (5) with 1.1 equivalents of phenyl chlorothionocarbonate (phenoxythiocarbonyl
chloride, PTC-Cl) and 2 equivalents of 4-dimethylaminopyridine
(DMAP) in acetonitrile generally resulted in clean formation of
the 2'-O-phenoxythiocarbonyl esters (6).   Relatively unhindered
secondary alcohols (e.g. cholesterol) react smoothly with
PTC-Cl in dichloromethane in the presence of 2-4 equivalents of
pyridine (12).   However, the use of DMAP/acetonitrile is required for more hindered systems including the 3',5'-O-TPDS-
nucleosides.   No protection of the heterocyclic amino groups
was necessary except with cytosine.   Selective N-acetylation of
cytidine (42) followed by its conversion to 4-N-acetyl-3',5'-O-
TPDS-cytidine proceeded smoothly.   Thioacylation of this compound was sluggish and incomplete under the usual conditions.
However, use of 6-9 equivalents of DMAP with the usual 10%
molar excess of PTC-Cl gave clean and rapid conversion to 4-N-
acetyl-2'-O-PTC-3',5'-O-TPDS-cytidine.

Smooth homolytic hydrogenolysis of the C2'—O2' bond of the
2'-phenylthionocarbonate esters (6) occurred using 1.5 equivalents of tri-n-butyltin hydride and 0.2 equivalents of AIBN
initiator at 70-80°C in deoxygenated toluene. Unless isolation
of the resulting 2'-deoxy-3',5'-O-TPDS-nucleoside (7) was
desired, in situ deprotection was effected directly by addition
of 2 equivalents of tetra-n-butylammonium fluoride (43).
Stirring of the warm toluene solution was continued for ~1
hour, and the desired 2'-deoxynucleoside (8) was isolated. The
overall conversion of a ribonucleoside (4) to its corresponding
2'-deoxynucleoside (8) could usually be performed without
purification of the intermediate products (5, 6, 7) beyond
partitioning, washing, and drying of the organic solution. The
product 2'-deoxynucleosides (8) were purified by column chromatography and/or recrystallization and were fully characterized. Overall yields refer to the analytially pure crystalline
products.

Adenosine was converted to 2'-deoxyadenosine in 78% yield
overall (12). This corresponds to average yields of 94% for
each of the 4 stages. The reactions generally proceed to
completion (as evaluated by thin layer chromatography) and
losses occurred during work-up, chromatography, and recrystallization. The corresponding overall yields for conversion of
the other major ribonucleosides to their 2'-deoxynucleosides
were 57% (guanosine), 68% (uridine), and 65% (cytidine).

As noted in the Introduction, tubercidin (1) had been
subjected to treatment with α-acetoxyisobutyryl halides (1,23)
and related 2',3'-acetoxonium ion mediated reactions (6).
However, only 3'-deoxytubercidin was isolated after hydrogenolysis of the resulting 3'-halo-3'-deoxy intermediates
(1,6,23). A multistep sequence that employed migration of a
benzylthio group from C3' to C2' gave 2'-deoxytubercidin in

~25% overall yield from the parent antibiotic (3). Enzymatic reductions of antibiotic nucleoside triphosphates also had been examined (44).

Treatment of tubercidin (1) and toyocamycin (7-cyano-7-deazaadenosine) by the presently described 4-stage sequence gave 2'-deoxytubercidin (68%) and 2'-deoxytoyocamycin (69%). Passage of an aqueous solution of the deprotected 2'-deoxy-toyocamycin through a column of Dowex 1-X2 (OH⁻) resin effected hydrolysis of the cyano function (45) to give 2'-deoxysangiv-amycin (2'-deoxy-7-carboxamido-7-deazaadenosine) in 65% yield overall from toyocamycin.

As expected, thiopurine nucleosides did not survive the reduction procedure. In order to expand the scope of our deoxygenation sequence, a chloropurine compound was exam-ined. Protection of 2-amino-6-chloro-9-(β-D-ribofuranosyl)-purine (see ref. 46 for an improved synthesis) to give its 3',5'-O-TPDS derivative was effected at 0°C to avoid dark coloration of the pyridine solution. Phenoxythiocarbonylation at O2' proceeded normally. Subjection of the resulting 2'-O-PTC-3',5'-O-TPDS 2-amino-6-chloropurine nucleoside to the usual reduction conditions at 75-80°C resulted in formation of multi-ple products (tlc). However, more gentle treatment of this compound with tri-n-butyltin hydride and AIBN in toluene at 65°C effected quite clean hydrogenolysis of the C2'—O2' bond. Deprotection and purification of the product using carbon followed by silica gel and recrystallization gave 2-amino-6-chloro-9-(2-deoxy-β-D-erythro-pentofuranosyl)purine in 56% overall yield. This compound can be converted to a variety of 2-amino-6-substituted purine deoxynucleosides by nucleo-philic replacement of the 6-chloro function (47).

Treatment of 9-(β-D-arabinofuranosyl)adenine with 1,3-dichloro-1,1,3,3-tetraisopropyldisiloxane/pyridine gave clean and exclusive formation of 3',5'-O-TPDS-araA. Conversion of

this compound to its 2'-O-PTC derivative proceeded smoothly.
Subjection of 2'-O-PTC-3',5'-O-TPDS-adenosine and 2'-O-PTC-
3',5'-O-TPDS-araA to the usual reduction conditions using tri-
n-butyltin deuteride followed by deprotection gave 2'-deuterio-
2'-deoxyadenosine with identical ratios (88:12) of ribo to
arabino-deuterio stereochemistry. This is in harmony with
preferential transfer of deuterium from the labeled stannane to
a C2' radical from the more sterically accessible α-face
(ribo). The 88% stereoselectivity provides a convenient biomimetic approach to the complete ribo selectivity executed by
ribonucleotide reductase (33).

III. A STEREODEFINED ROUTE TO α-2'-DEOXYNUCLEOSIDES

Condensation of an activated 2-deoxy sugar derivative with
a nucleobase is well known to give α,β-anomeric mixtures owing
to the absence of a participating group at C2. Preferential
anomeric ratios can be manipulated in some cases, but no
general stereoselective methods have been described previously.
Coupling of 1,2,3,5-tetra-O-acetyl-D-arabinofuranose (9)
(48) (see Figure 3) with bases using the general procedure of
Vorbrüggen (49) gave the expected α-D-arabinofuranosyl compounds in harmony with the well-known "trans rule" effect of
the β-oriented 2-O-acetyl group. Deacetylation followed by
protection of the resulting α-arabinonucleosides (10) under the
usual conditions gave the 3',5'-O-TPDS compounds (11) cleanly. Functionalization of O2' using PTC-Cl/DMAP/acetonitrile
proceeded smoothly. Homolytic hydrogenolysis of the 2'-O-PTC-
3',5'-O-TPDS-α-D-arabinonucleoside compounds followed by
in situ deprotection using tetra-n-butylammonium fluoride
provided a general stereodefined route to α-2'-deoxynucleosides
(12). Examples involving the common bases (adenine, guanine,

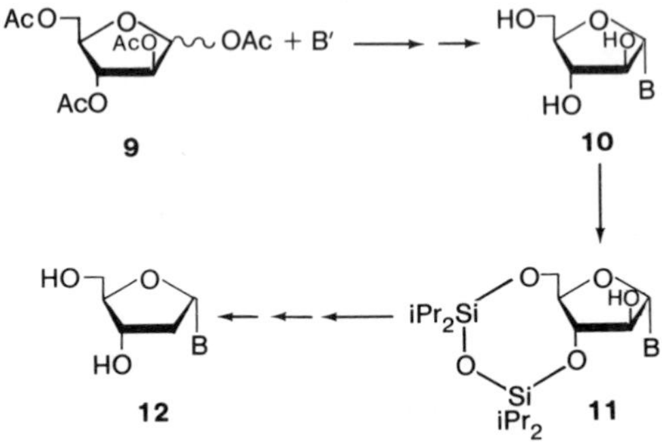

FIGURE 3

cytosine, and uracil) and synthesis of the anomerically pure 2-amino-6-chloro-9-(2-deoxy-α-D-<u>erythro</u>-pentofuranosyl)purine for conversion to the selective antineoplastic agent (36) "α-2'-deoxythioguanosine" have been completed. Examination of a more direct approach is in progress.

IV. AN OX-RED CONVERSION OF RIBO TO ARABINONUCLEOSIDES

Moffatt and co-workers had examined oxidation of di-protected nucleosides to give ketonucleosides and had reduced 2'-keto intermediates to give the arabinonucleosides of uracil and cytosine (50,51). However, the difficulty in obtaining the selectively protected precursors and the reported instability of the ketonucleosides have resulted in their being regarded as rather inaccessible compounds.

We have investigated oxidation of 3',5'-<u>O</u>-TPDS-nucleosides (<u>5</u>) (see Figure 4) followed by reduction of the 2'-keto inter-mediates (<u>13</u>) to give the arabinonucleosides (<u>14</u>) as major products after deprotection. A Moffatt-type oxidation using acetic anhydride/dimethylsulfoxide (52) was employed with most nucleosides bearing an amino group on the base. Chromium

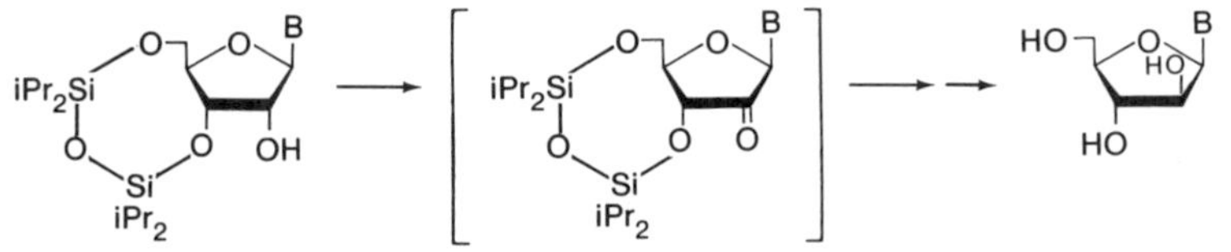

5   13   14

FIGURE 4

trioxide/pyridine/acetic anhydride (53) was found to be a
superior oxidant with less basic nucleosides. Reduction of the
2′-keto function was effected using lithium triethylborohyride
or sodium borohydride. The latter reagent works effectively
and was used routinely. Exploratory reductions using sodium
borodeuteride provided a method for evaluating the stereoselectivity of the reduction step. Measurement of the ratios
of 2′-protio:2′-deuterio ribonucleosides by NMR spectroscopy
allowed analysis of the quantity of unoxidized starting
material carried through the sequence. The ratio of 2′-
deuterioarabino:2′-deuterioribo nucleosides determined the
stereoselectivity.

   Conversion of adenosine, inosine, 2-aminoadenosine, and
guanosine as well as the nucleoside antibiotics tubercidin and
toyocamycin (with subsequent conversion to the sangivamycin
analogue) to their arabino epimers has been completed.
Pyrimidine nucleosides also are subject to this sequence, and
2′-labeled arabinonucleosides can be prepared with hydrogen
isotopes. However, the O2→C2′ cyclonucleoside approach (19) is
more practical for routine preparation of pyrimidine arabinonucleosides. A more extensive study of the synthesis and
reduction of ketonucleosides is in progress and will be
reported separately.

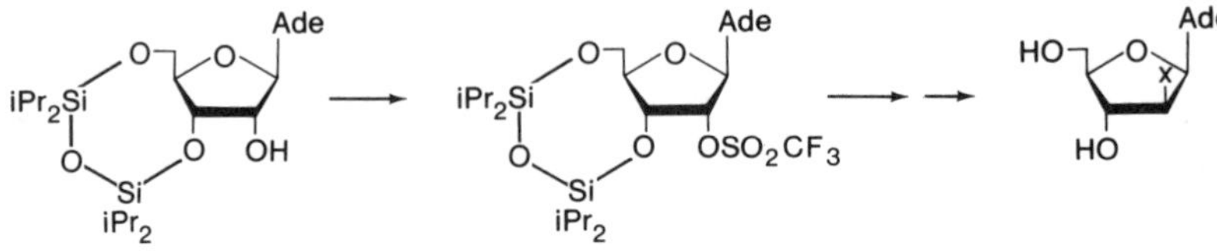

FIGURE 5

V. SYNTHESES OF 2'-SUBSTITUTED-2'-DEOXY ARABINO
AND RIBONUCLEOSIDES

Activation of O2' of purine nucleosides by trifluoro-
methylsulfonylation allows nucleophilic displacement of the
resulting 2'-triflate group with inversion of configuration.
However, multistep sequences were required previously to obtain
the desired protected nucleoside precursors (18,27,28).
Specific triflation conditions also are crucial for achieving
efficient overall yields.

Treatment of 3',5'-O-TPDS-adenosine (15) (see Figure 5)
with trifluoromethanesulfonyl chloride in dichloromethane
solution at 0°C in the presence of 3 equivalents of DMAP gave
2'-O-triflyl-3',5'-O-TPDS-adenosine (16) in 74% yield as a pure
crystalline product. Treatment of (16) with azide, bromide,
chloride, and iodide nucleophiles followed by the usual deprotection gave the respective 9-(2-substituted-2-deoxy-β-D-
arabinofuranosyl)adenine products (17). Hydrogenolysis of the
2'-azido compound provided the known (8) 2'-amino-2'-deoxy-araA
(17, X = NH$_2$) in ~55% overall yield from (15).

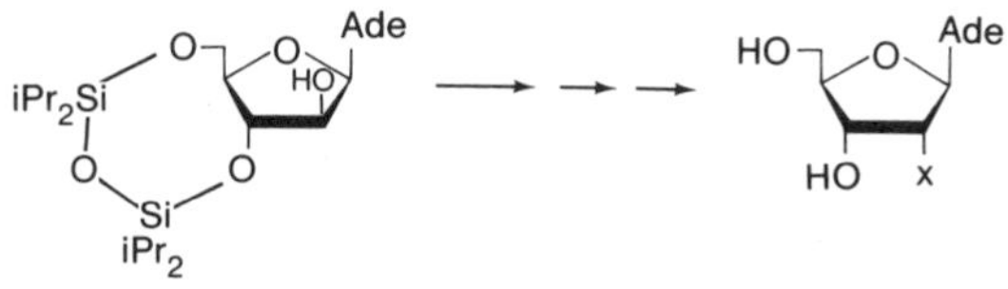

18 **19**

FIGURE 6

An analogous sequence beginning with triflation of 3',5'-O-TPDS-araA (18) (see Figure 6) followed by nucleophilic displacements and deprotection gave the corresponding 2'-substituted-2'-deoxyadenosine (19) compounds (18,28). As expected, displacements of the arabino triflate at the less hindered α-face to give the ribo products (19) proceeded more readily. Hydrogenolysis of the 2'-azido compound gave the nucleoside antibiotic (54) 2'-amino-2'-deoxyadenosine (19, X = NH$_2$) in ~45% overall yield.

VI. CONCLUSIONS

The Markiewicz-Wiewiórowski protection of nucleosides as their 3',5'-O-TPDS derivatives and development of reactions that proceed at the 2'-position in high yields have rendered the previously difficult modification of C2' (especially with purine-type nucleosides) readily accessible. Treatment of arabino and ribonucleosides with 1,3-dichloro-1,1,3,3-tetraisopropyldisiloxane in pyridine generally gives high yields of the selectively di-protected 3',5'-O-TPDS compounds. Phenoxythiocarbonylation at O2' followed by homolytic hydrogenolysis of the C2'—O2' linkage of these phenyl thionocarbonate esters and disiloxyl deprotection provides a 4-stage high-yield conversion of nucleosides to 2'-deoxynucleosides. Biomimetic stereoselectivity of 88% for delivery of deuterium to the α-face was observed in the synthesis of 2'-deuterio-2'-deoxyadenosine. Coupling of 1,2,3,5-tetra-O-acetyl-D-arabino-furanose with nucleobases proceeds as expected to give the α-arabinonucleosides after deacetylation. Subjection of these anomerically defined compounds to the 4-stage conversion sequence provides the first general route to 2'-deoxy-α-D-erythro-pentofuranosyl nucleosides. Trifluoromethanesulfonylation of adenine 3',5'-O-TPDS-nucleosides followed by nucleophilic displacements of the 2'-triflate group and deprotection affords 2'-substituted-2'-deoxy-β-D-arabino or ribo nucleosides from their ribo or arabino precursors, respectively. Experimental details for these and related transformations will be provided in journal publications.

REFERENCES

1. M. J. Robins, J. R. McCarthy, Jr., R. A. Jones, and R. Mengel. Can. J. Chem., 51, 1313 (1973).
2. M. J. Robins, Y. Fouron, and R. Mengel. J. Org. Chem., 39, 1564 (1974).
3. M. J. Robins and W. H. Muhs. J. Chem. Soc., Chem. Commun., 269 (1976).
4. M. J. Robins, R. Mengel, R. A. Jones, and Y. Fouron. J. Am. Chem. Soc., 98, 8204 (1976).
5. M. J. Robins, R. A. Jones, and R. Mengel. J. Am. Chem. Soc., 98, 8213 (1976).
6. M. J. Robins, R. A. Jones, and R. Mengel. Can. J. Chem., 55, 1251 (1977).
7. M. J. Robins and W. H. Muhs. J. Chem. Soc., Chem. Commun., 677 (1978).
8. M. J. Robins and S. D. Hawrelak. Tetrahedron Lett., 3653 (1978).

9. M. J. Robins. in "Nucleosides, Nucleotides and Their Biological Applications", INSERM Colloque 1978, Vol. 81, J.-L. Barascut and J.-L. Imbach, Eds., INSERM, Paris, 1979, pp. 13-35.

10. M. J. Robins, P. Sporns, and W. H. Muhs. Can. J. Chem., 57, 274 (1979).

11. M. J. Robins, S. D. Hawrelak, T. Kanai, J.-M. Seifert, and R. Mengel. J. Org. Chem., 44, 1317 (1979).

12. M. J. Robins and J. S. Wilson. J. Am. Chem. Soc., 103, 932 (1981).

13. D. M. Brown, D. B. Parihar, C. B. Reese, and Sir A. Todd. J. Chem. Soc., 3035 (1958).

14. J. F. Codington, I. L. Doerr, and J. J. Fox. J. Org. Chem., 29, 558; 564 (1964).

15. A. Holý. Collect. Czech. Chem. Commun., 37, 4072 (1972).

16. M. Ikehara and H. Tada. Chem. Pharm. Bull, 15, 94 (1967).

17. M. Ikehara, T. Maruyama, H. Miki, and Y. Takatsuka. Chem. Pharm. Bull., 25, 754 (1977).

18. M. Ikehara, T. Maruyama, and H. Miki. Tetrahedron, 34, 1133 (1978).

19. J. G. Moffatt. in "Nucleoside Analogues: Chemistry, Biology, and Medical Applications", R. T. Walker, E. De Clercq, and F. Eckstein, Eds., Plenum Press, New York, 1979, pp. 71-164.

20. W. T. Markiewicz and M. Wiewiórowski. Nucleic Acids Res., Spec. Publ. No. 4, s185 (1978).

21. W. T. Markiewicz. J. Chem. Res. (S), 24 (1979).

22. A. F. Russell, S. Greenberg, and J. G. Moffatt. J. Am. Chem. Soc., 95, 4025 (1973).

23. T. C. Jain, A. F. Russell, and J. G. Moffatt. J. Org. Chem., 38, 3179 (1973).

24. J. Abola and M. Sundaralingam. Acta Crystallogr., B29, 697 (1973).

25. T. F. Lai and R. E. Marsh. Acta Crystallogr., B28, 1982
    (1972).

26. P. Prusiner, T. Brennan, and M. Sundaralingam.
    Biochemistry, 12, 1196 (1973).

27. R. Ranganathan and D. Larwood. Tetrahedron Lett., 4341
    (1978).

28. M. Ikehara and H. Miki. Chem. Pharm. Bull., 26, 2449
    (1978).

29. R. E. Ireland, D. C. Muchmore, and U. Hengartner. J. Am.
    Chem. Soc., 94, 5098 (1972).

30. R. E. Ireland, C. S. Wilcox, and S. Thaisrivongs. J. Org.
    Chem., 43, 786 (1978).

31. T. C. Jain, I. D. Jenkins, A. F. Russell, J. P. H.
    Verheyden, and J. G. Moffatt. J. Org. Chem., 39, 30
    (1974).

32. T. Adachi, T. Iwasaki, I. Inoue, and M. Miyoshi. J. Org.
    Chem., 44, 1404 (1979).

33. L. Thelander and P. Reichard. Annu. Rev. Biochem., 48,
    133 (1979).

34. D. H. R. Barton and S. W. McCombie. J. Chem. Soc., Perkin
    Trans. I, 1574 (1975).

35. D. H. R. Barton and R. Subramanian. J. Chem Soc., Perkin
    Trans. I, 1718 (1977).

36. E. M. Acton, R. N. Goerner, H. S. Uh, K. J. Ryan, D. W.
    Henry, C. E. Cass, and G. A. LePage. J. Med. Chem., 22,
    518 (1979).

37. K. Fukukawa, T. Ueda, and T. Hirano. Chem. Pharm. Bull.,
    29, 597 (1981).

38. R. A. Lessor and N. J. Leonard. J. Org. Chem., 46, 4300
    (1981).

39. K. Pankiewicz, A. Matsuda, and K. A. Watanabe. J. Org.
    Chem., 47, 485 (1982).

40. C. H. M. Verdegaal, P. L. Jansse, J. F. M. de Rooij, and J. H. van Boom. Tetrahedron Lett., 21, 1571 (1980).

41. W. T. Markiewicz and M. Wiewiórowski. Nucleic Acids Res., Symp. Ser. No. 11, 21 (1982).

42. K. A. Watanabe and J. J. Fox. Angew. Chem., Int. Ed. Engl., 5, 579 (1966).

43. E. J. Corey and A. Venkateswarlu. J. Am. Chem. Soc., 94, 6190 (1972).

44. S. A. Brinkley, A. Lewis, W. J. Critz, L. L. Witt, L. B. Townsend, and R. L. Blakley. Biochemistry, 17, 2350 (1978).

45. T. Uematsu and R. J. Suhadolnik. Biochemistry, 9, 1260 (1970).

46. M. J. Robins and B. Uznański. Can. J. Chem., 59, 2601 (1981).

47. R. H. Iwamoto, E. M. Acton, and L. Goodman. J. Med. Chem., 6, 684 (1963).

48. B. L. Kam, J.-L. Barascut, and J.-L. Imbach. Carbohydr. Res., 69, 135 (1979).

49. H. Vorbrüggen, K. Krolikiewicz, and B. Bennua. Chem. Ber., 114, 1234 (1981).

50. A. F. Cook and J. G. Moffatt. J. Am. Chem. Soc., 89, 2697 (1967).

51. U. Brodbeck and J. G. Moffatt. J. Org. Chem. 35, 3552 (1970).

52. J. D. Albright and L. Goldman. J. Am. Chem. Soc., 89, 2416 (1967).

53. P. J. Garegg and B. Samuelsson. Carbohydr. Res., 67, 267 (1978).

54. Y. Iwai, A. Nakagawa, and A. Nagai. J. Antibiot., 32, 1367 (1979).

SYNTHESIS OF VERSATILE C-NUCLEOSIDE PRECURSORS AND CERTAIN C-NUCLEOSIDES[1]

Dean S. Wise
Thomas L. Cupps
John C. Krauss
Leroy B. Townsend

Department of Medicinal Chemistry
College of Pharmacy
Department of Chemistry
University of Michigan
Ann Arbor, Michigan 48109

The biological importance of many naturally occurring C-nucleosides has prompted considerable interest in the exploration of synthetic routes to produce not only these compounds but also structurally related analogs. Recent reviews[1] have described the numerous strategies which have been developed for the successful synthesis of C-nucleosides, such as, oxazinomycin, showdomycin, pyrazomycin, pyrazomycin B, formycin, formycin B, oxoformycin B, as well as several structural congeners.

Of these nucleosides, formycin (1) has been of particular interest. At least three different strategies leading to the synthesis of formycin[2] (1) have been reported[3-5] as a result of its demonstrated chemotherapeutic and biological activity as an adenosine congener. Formycin (1) has shown antitumor, antibacterial, antifungal and antiviral activity.[6] This naturally occurring compound is converted to its 5'-monophosphate by the enzyme adenosine kinase.[7] Formycin is also a substrate for the enzyme adenosine deaminase[8] (ADA) and

[1]This study was supported by funds from the National Institutes of Health (Training Grant No. 5-T32-GM 07767) and the American Cancer Society (Grant CH-133).

is thereby catabolized to the inosine analog formycin B ($\underline{2}$), which demonstrates a much lower level of chemotherapeutic activity. Hepatic aldehyde oxidase[9] further converts formycin B ($\underline{2}$) to the biologically inactive xanthosine analog oxoformycin B ($\underline{3}$). The facility with which formycin ($\underline{1}$) is catabolized[10]

1   2   3

in human tissue starting with ADA has severely limited its therapeutic potential and suggests that research designed to afford a structural analog of $\underline{1}$ might result in the development of a compound which would not be a substrate for ADA but would still possess significant biological activity. One approach to the construction of analogs of formycin could involve subtle changes in the pyrazolo[4,3-$\underline{d}$]pyrimidine heterocyclic ring system.

Recently, the total synthesis of just such a congener of formycin, 9-deazaadenosine[11] ($\underline{4}$) as well as that of 9-deazainosine[12] ($\underline{5}$) was reported and preliminary <u>in vitro</u> data, indeed, indicated that $\underline{4}$ has pronounced growth inhibitory activity against several mouse and human cell lines.[13] This suggests that $\underline{4}$ is resistant to ADA and most likely functions as a substrate for adenosine kinase. The activity reported for $\underline{4}$

lends further credence to the importance of preparing additional adenosine C-nucleoside congeners. It should be mentioned that the first synthesis of a 7-ribosylpyrrolo[3,2-d]pyrimidine was reported[14] in 1976. In this report the investigators described the synthesis of 9-deazaxanthosine (6).

It is apparent that the published methods for the synthesis of compounds 4 - 6, as well as formycin all follow basically the same strategy. Each method starts from a suitably derivatized sugar and then proceeds through a step by step construction of

the five-membered heterocyclic ring followed by an annulation to afford the bicyclic aglycon attached by a carbon-carbon bond to the sugar. It is worth noting that each method represents a minimum of 8 synthetic steps from a suitably blocked ribose derivative.

The investigations presented in this talk were undertaken in order to develop versatile and shorter pathways to C-nucleosides. We were interested in the development of new precursors, and new methodologies which could be adapted to the synthesis of several classes of C-nucleosides. One of the

targets in this general scheme and toward which we will thread
our way, is 2,4-dimethoxy-7-(ß-D-ribofuranosyl)pyrrolo[3,2-d]-
pyrimidine (7). Not only is the synthesis of 7 undocumented,
but the 2,4-dialkoxypyrrolo[3,2-d]pyrimidine ring system is
amenable for ready modification.

The general scheme (Scheme I) we envisaged for the
synthesis of compound 7 is divided into two parts which begin
with   with   1-O-acetyl-2,3,5-tri-O-benzoyl-ß-D-ribofuranose[15]
(8) and 6-methyluracil (9), respectively. Although there is a
large structural difference in the starting compounds, both
pathways include as crucial intermediates the structurally
related 6-C-(1-ß-D-ribofuranosyl)methylpyrimidines 11 and 12.
C-Nucleosides of this type have been referred to as
"homo-C-nucleosides."[16]   The two paths finally converge
through pyrimidine intermediates with the synthesis of the
target compound 7.

7

Aside from the pyrimidine-then-pyrrole order of ring
formation, path A still resembles the published synthetic
routes[11,12] to bicyclic C-nucleosides, since it begins with a
protected glycoside and follows through a step by step
construction of the heterocyclic moiety. However, there are

SCHEME I

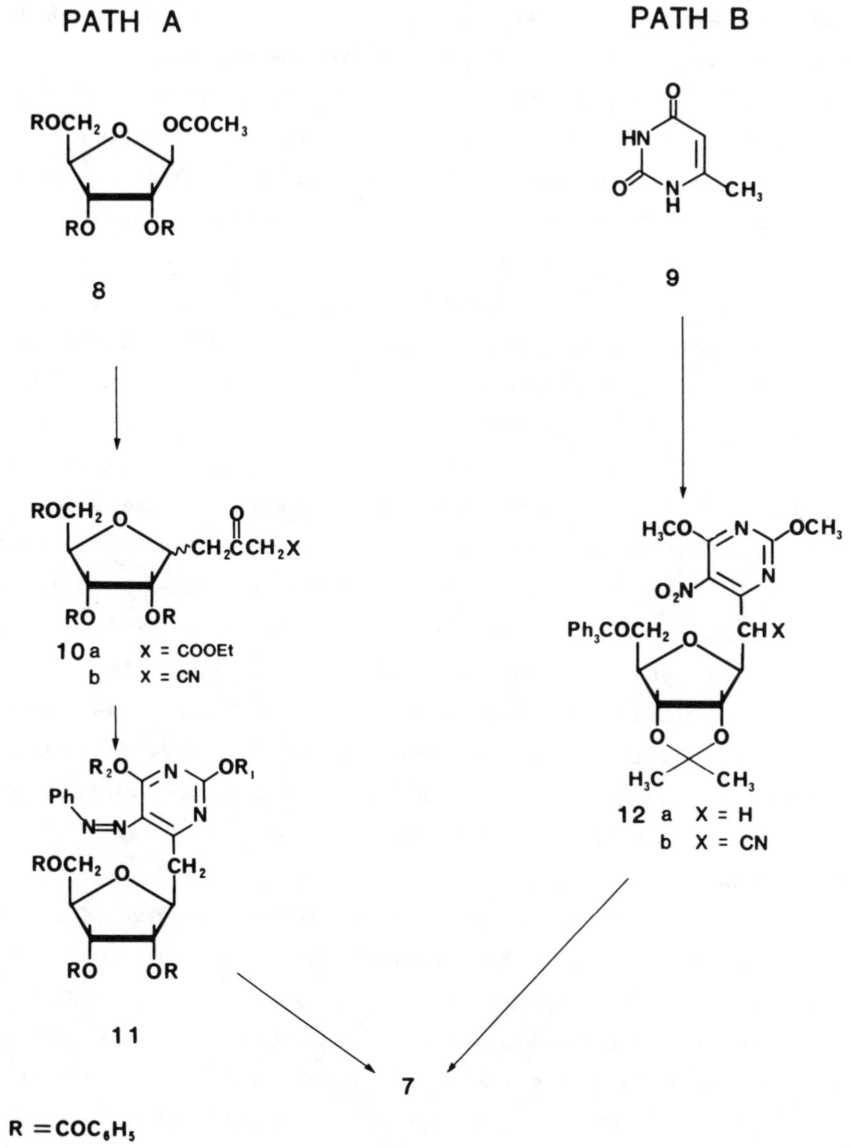
PATH A
PATH B
ROCH2
OCOCH3
RO
OR
8
HN
O
N
H
CH3
9
CH2CCH2X
10 a X = COOEt
b X = CN
H3CO
N
OCH3
O2N
Ph3COCH2
CHX
H3C
CH3
12 a X = H
b X = CN
R2O
OR1
Ph
N=N
CH2
11
7
R =COC6H5

several points in Path A (Scheme I) which make this an attractive approach. First the starting sugar 8 is crystalline and inexpensive, and contains a 2'-benzoyloxy group which has been previously shown to provide anchimeric assistance[17] in Lewis acid catalyzed condensations involving nucleophiles and the anomeric center of the sugar. It has been established that these condensations lead to the formation of the ß anomer in the case of N-nucleophiles. Some cases involving C-nucleophiles have been reported[18] to be under the same stereo-control, e.g. a reaction between the trimethylsilyl enol ether of cyclo-hexanone and 8 has been reported to produce ß-ribosylated cyclohexanone in high yield. Another interesting point in Path A lies in the chemical nature of compound 10. Since ß-ketoesters[19] as well as ß-ketonitriles[20] have long represented classical synthetic intermediates for the synthesis of pyrimidine heterocycles, compounds 10a and 10b should provide access to substituted pyrimidines of the general structure 11 when reacted with dinucleophiles such as guanidine, thiourea and urea. The conversion of substituted pyrimidines into pyrrolo-[3,2-d]pyrimidines has been reviewed.[21] Finally, ß-ketoesters are known to readily add diazonium salts[22] to form phenyldiazo esters which may be converted[23] to N-phenylpyridazinones and pyrrolo[3,2-d]pyrimidines via an intermediate pyrimido[4,5-c]-pyridazine.

A number of procedures[1] have been developed for the condensation of carbon nucleophiles with the anomeric carbon atom of suitable carbohydrates. In one such procedure,[24] 1-O-acetyl-2,3,5-tri-O-benzoyl-ß-D-ribofuranose (8) was reacted with 1-hexene in the presence of stannic chloride to afford a good yield (75%) of 1-(2,3,5-tri-O-benzoyl-D-ribofuranosyl)-2-hexene,[25] 13a and (13b)[26] (unspecified anomeric configuration). Through the oxidation of the olefinic bond of

13, one can envision the formation of a potentially useful C-nucleoside precursor. Indeed, when the authors[24] treated compound 13a(13b) with $KMnO_4$ and KIO_4, 2-(2,3,5-tri-O-benzoyl-D-ribofuranosyl)acetic acid (14a) was formed in 32% yield. This compound (14a) was subsequently converted to the ethyl ester 14b.

8

13a V=H, W = $CH_2CH=CHC_3H_7$
  b V=$CH_2CH=CHC_3H_7$, W=H

14a R = H
    R = Et

Other compounds, structurally similar to 14b, containing an active methylene group attached to the anomeric carbon also have been shown to be good synthetic intermediates in the synthesis of several important C-nucleosides, e.g. the successful synthesis[27] of several pyrimidine C-nucleosides from the intermediate 15. Also, the syntheses of oxazinomycin,[28] 9-deazaadenosine,[29] 9-deazainosine,[30] and several analogs of pseudouridine[31] have recently been accomplished starting from the protected sugars, 16a and 16b. In view of the above, we felt that the condensation of a per-O-acylglycoside with 1-hexene warranted a closer investigation to determine its feasibility as an economical route for the formation of an important intermediate in C-nucleoside synthesis.

Unfortunately, the reaction of 1-hexene with either 1-O-acetyl-2,3,5-tri-O-benzoyl-ß-D-ribofuranose (8) or 1,2,3,5-tetra-O-acetyl-ß-D-ribofuranose (8a) in the presence of stannic chloride leads to the formation of a complex mixture of products (See Figures 3, 4). By a combination of ^{1}H-nmr and mass spectroscopy, the structures of the products were shown to be anomeric and diastereomeric mixtures of the 8,9,11-tri-O-acylprotected derivatives of 7,10-anhydro-5-chloro-1,2,3,4,5,6-hexadeoxy-D-allo(altro)-undec-4-enitol (13) and 7,10-anhydro-5-chloro-1,2,3,4,5,6-hexadeoxy-D-allo(altro)-undecitol (17). (See Figure 3). The α anomer of 13 was the predominant anomer while

15

16a   X=OH, R=Tr

16b   X=OH, R=H

16c   X=Cl , R=Tr

the α and ß anomers of 17 were present in approximately equal amounts. It was found that the formation of 17 could be precluded when trimethylsilyl trifluoromethanesulfonate was used as the catalyst instead of stannic chloride. The acyl protected sugar 3,6-anhydro-2-deoxy-D-allo(altro)-heptose (18), prepared by ozonolysis of 13, reacted with t-butoxycarbonylmethyl-triphenylphosphorane to give t-butyl trans-5,8-anhydro-6,7,9-tri-O-acetyl-2,3,4-trideoxy-D-allo-(altro)-non-2-enonate (19). The basicity of the ylide was sufficient to cause anomerization. and resulted in an α/ß ratio of 5/1 in the product 19.

The fact that 1-hexene possessed sufficient nucleophilicity to react at all suggested that perhaps better and more efficient "soft" nucleophiles could be found. Our search for such an alternative reagent led us to examine the use of allyltrimethylsilane (allyl-TMS) as the condensing agent in the

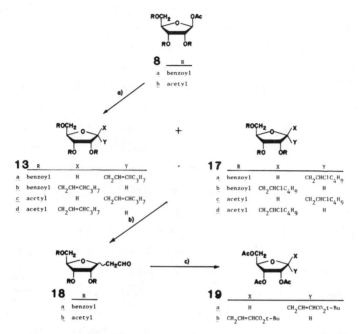

a) 1-hexene, SnCl$_4$, CH$_3$CN, rt.   b) O$_3$, methanol, -78°, dimethylsulfide.   c) t-butoxycarbonyl-methyltriphenylphosphorane, CH$_3$CN, rt.

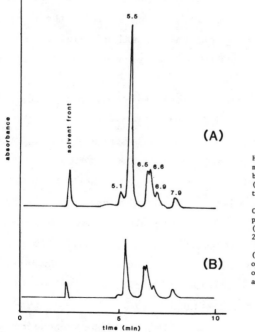

(A)

(B)

HPLC chromatograms of reaction mixtures of (A) the reaction between **8** and 1-hexene, and (B) the ozonolysis reaction on the mixture (A).

Column: Micropak Si-5; liquid phase: hexane:ethyl acetate (85:15, v/v); uv detector: 254 nm.

(Note: The product of the ozonolysis, the resulting ozonide, elutes only after about 30 min.).

above reaction.

The mildly nucleophilic nature of allyl-TMS has been demonstrated,[33] particularly in the presence of Lewis acids such as $TiCl_4$, BF_3, or $AlCl_3$. Also of considerable interest was a recent account[34] which described the reaction of allyl-TMS with acetals to form substituted allyl ethers in the presence of the Lewis acid, trimethylsilyl trifluoromethane-sulfonate (TMS triflate). Since the anomeric centers in tetrasubstituted pentose sugars are indeed acetals, it seemed reasonable to assume that suitably substituted sugars might also react under similar conditions with allyl-TMS to form allyl sugar derivatives of the general form shown in compound 20.

The reaction of allyl-TMS with 1-O-acetyl-2,3,5-tri-O-benzoyl-ß-D-ribofuranose,[35] in the presence of excess TMS triflate, formed 4,7-anhydro-1,2,3-trideoxy-5,6,8-tri-O-benzoyl-D-altro-oct-1-enitol (20a) and 4,7-anhydro-1,2,3-trideoxy-5,6,8-tri-O-benzoyl-D-allo-oct-1-enitol (20b) in a combined yield of 83% and in a 10/1 ratio respectively (see SCHEME 2). That 20a and 20b were anomers was suggested by their similar [1]H-nmr data but differing optical rotation values. However, the unequivocal proof of structure was provided by the following chemical transformations.

Compound 20 was readily oxidized with ozone to afford 3,6-anhydro-2-deoxy-4,5,7-tri-O-benzoyl-D-heptose (18) or 3,6-anhydro-2-deoxy-4,5,7-tri-O-benzoyl-D-hepanoic acid (14a). Few problems were encountered in the synthesis of the ß-ketonitrile 23[36] or the ß-ketoester 25. Both intermediates were obtained in high yield through the carboxylic acid 14. Exploratory reactions to form the pyrimidine homo-C-nucleosides, by ring closure using reagents such as urea, thiourea and guanidine, were for the most part unsuccessful. However, the reaction of 25 with guanidine did afford 2-amino-6-C-(1-D-ribofuranosyl)methylprimidin-4-one[16] (26). It was found that compound 23 or 25 would react readily with benzenediazonium

SCHEME 2

R = -COPh

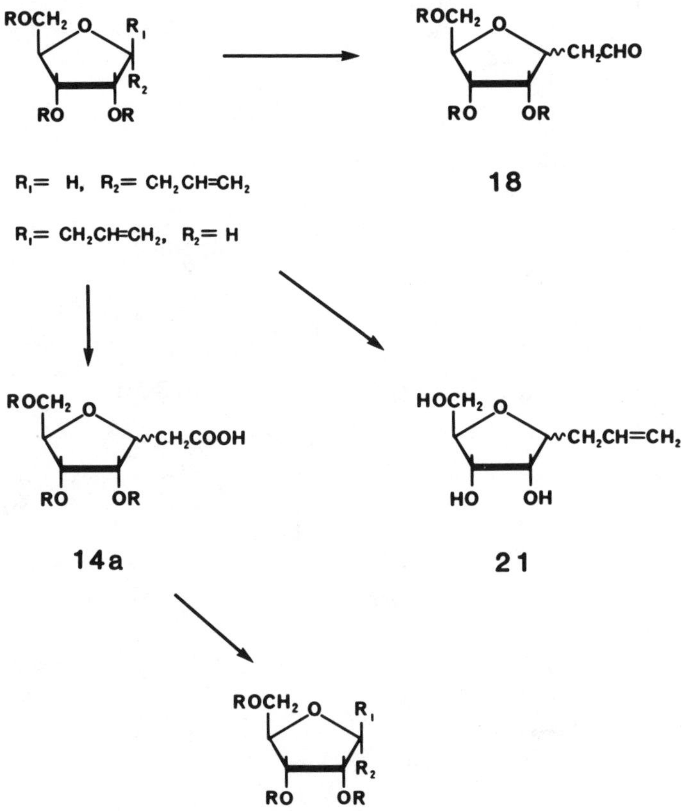

20 a R_1= H, R_2= $CH_2CH{=}CH_2$

b R_1= $CH_2CH{=}CH_2$, R_2= H

18

14a

21

14b R_1=H, R_2= CH_2COOEt

R_1= CH_2COOEt, R_2= H

SCHEME 3

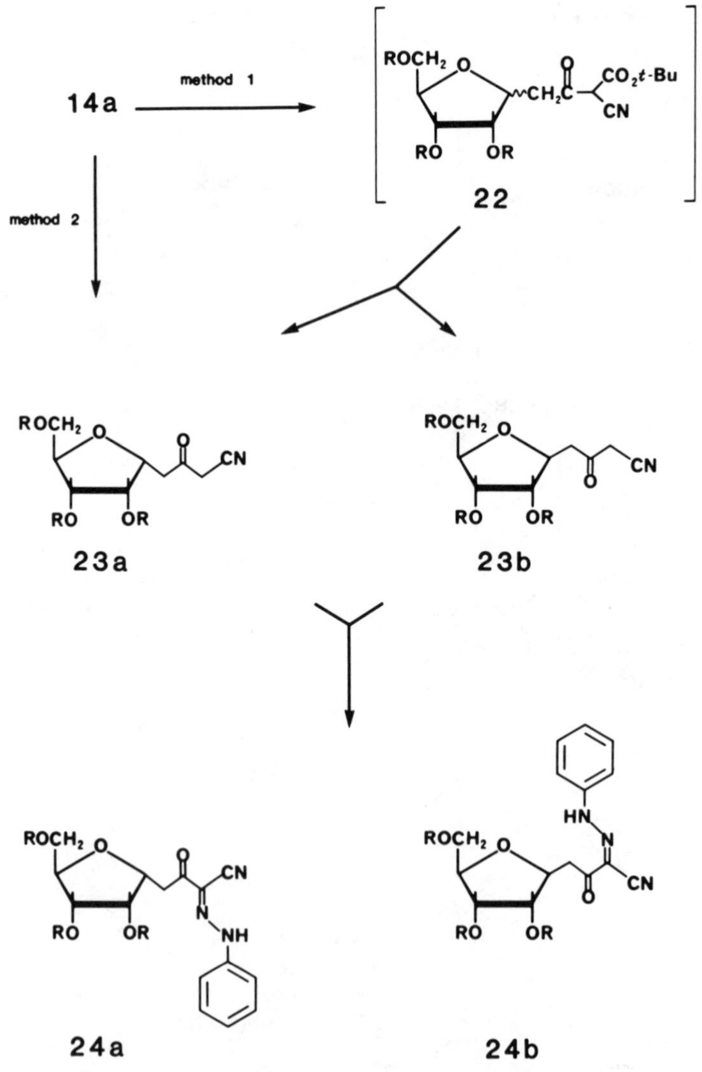

chloride to form the phenyl-diazoketones 24 or 27, 28 in yields
ranging from 90% to 100%. However, in spite of numerous
attempts, 24 and 27, 28 could not be converted into the
5-phenyldiazopyrimidine.

It is perhaps worthwhile to list at this time the major
drawbacks which prevented the synthesis of 7 by the reactions
shown in Scheme 1 Path A. First, throughout the scheme, the
problem of anomeric preference was never resolved in favor of
the ß anomer. Although one of the procedures which was
developed for the synthesis of the ß-ketonitrile 23 preferentially yielded the ß epimer, the same preference was not observed
for the reaction involving the synthesis of the ß-ketoester 25.

Another problem which plagued Scheme 1 Path A was the base
labile benzoyl protecting groups which were present in the
starting sugar 8. Although this type of group has been shown to
lend anchimeric assistance[19] for the formation of ß glycosidic
bonds, this was obviously not the case in the present example,
since the condensation of 8 with allyl-TMS proceeded with a
predomination of the α anomer. In addition, the benzoyl groups
were later found to be problematic in the pyrimidine forming
reactions involving thiourea and urea. These problems were
presumably due to the basic conditions which are necessary to
form the reactive anion. These conditions were apparently
severe enough to effect a removal of the benzoyl groups, a
competitive process which seemed to interfere with the desired
ring closure. Another problem associated with Scheme 1 Path A
was the fact that it failed to furnish the desired pyrimidine
homo-C-nucleoside 11. This was supported by the failure to
convert the phenyl-diazoketone 27 into 11 in spite of the fact
that a similar reaction[30] had been successful in the case of a
non-ribosylated ketone.

SCHEME 4

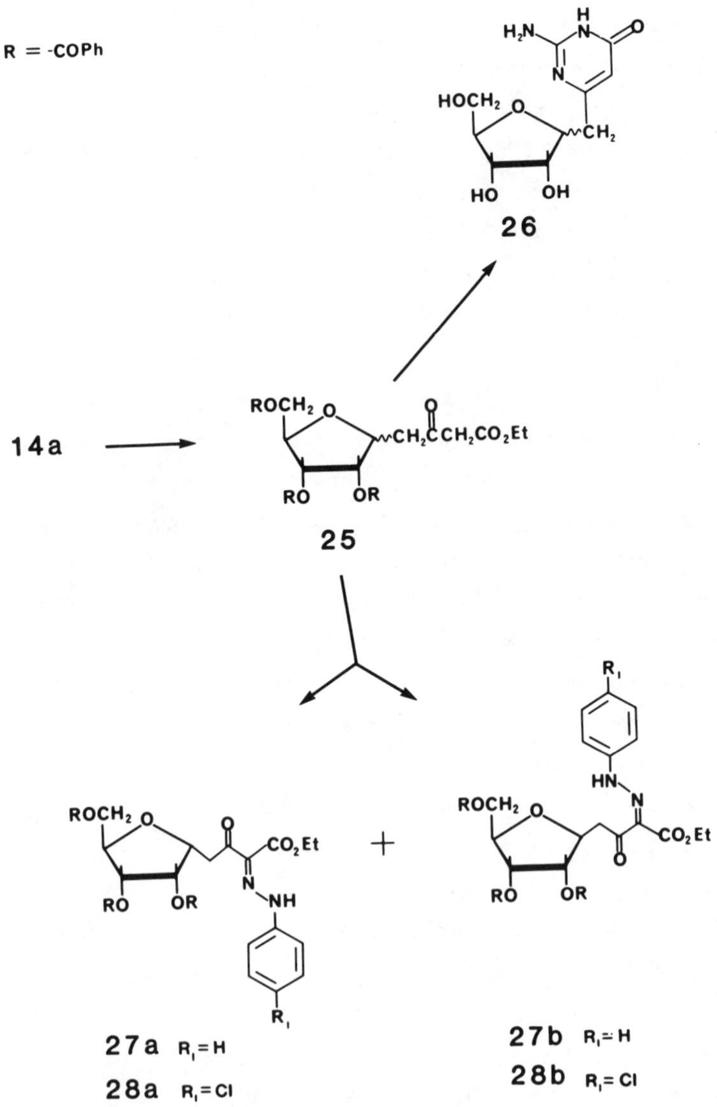

R = -COPh

14a $\longrightarrow$

25

27a $R_1 = H$
28a $R_1 = Cl$

27b $R_1 = H$
28b $R_1 = Cl$

Fortunately, by the time these major difficulties in Scheme 1 Path A were encountered, an investigation of an alternate route (Scheme 1, Path B) by which compound 7 could be synthesized already 14 had been initiated. This second route involved the synthesis of 7 starting with 6-methyluracil (9) and preceeded via one of two condensation reactions (Scheme 5) to form the target intermediate compound 12. This approach was successful in avoiding the pitfalls uncovered in Scheme 1 Path A and led, indeed, to the eventual synthesis of compound 7.

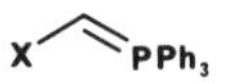

X=CO₂Et
X=CN

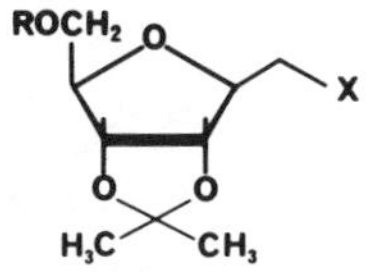

29 R = Tr, X = CO₂Et
30 R = Tr, X = CN
31 R = H , X = CO₂Et

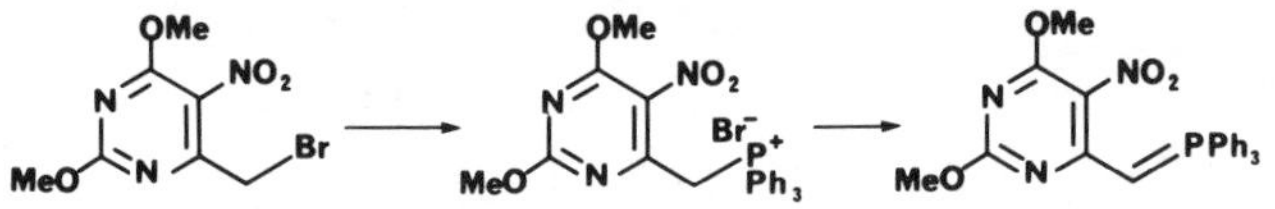

32   33   34

Me = CH₃

As was stated above, the crucial step in this reaction scheme involved one of two condensation procedures outlined in Scheme 5. The condensation of stabilized ylides such as

(carboethoxymethylene)triphenylphosphorane or (cyanomethylene)-
triphenylphosphorane with 2,3-$\underline{O}$-isopropylidene-5-$\underline{O}$-triphenyl-
methyl-ß-D-ribofuranose[37] or 2,3-isopropylidene-$\underline{D}$-ribofuran-
ose[37,38] has afforded the versatile $\underline{C}$-nucleoside precursors
ethyl 2-(2,3-$\underline{O}$-isopropylidene-t-$\underline{O}$-triphenylmethyl-1-ß-$\underline{D}$-ribo-
furanosyl)acetate (<u>29</u>), and 2-(2,3-$\underline{O}$-isopropylidene-5-$\underline{O}$-
triphenylmethyl-1-ß-$\underline{D}$-ribofuranosyl)acetonitrile (<u>30</u>), ethyl
2-(2,3-$\underline{O}$-isopropylidene-1-ß-$\underline{D}$-ribofuranosyl)acetate (<u>31a</u>) or
2-(2,3-$\underline{O}$-isopropylidene-5-$\underline{O}$-triphenylmethyl(1-ß-$\underline{D}$-ribofura-
nosyl)acetonitrile (<u>31b</u>), respectively. In spite of the fact
that compounds <u>29</u> and <u>30</u> have played key roles in the
synthesis[1] of a variety of biologically active $\underline{C}$-nucleosides,
it was somewhat surprising to see on reviewing the literature
that no work had been done to extend the scope and versatility
of this sugar:ylide condensation reaction and that an obvious
extension of this procedure would be to carry out the
condensation of either of these sugars with a different
stablized ylide. It occurred to us that one such alternate
phosphorous ylide might be one in which the electron-withdrawing
group is a suitably substituted pyrimidine.

It was felt that if a stable, isolable pyrimidine ylide
were to be synthesized, it would have to contain a more electron
deficient pyrimidine system such as would be the case in a
5-nitropyrimidine. We, therefore, chose to prepare as a target
the 6-$\underline{C}$-(1-$\underline{D}$-ribofuranosyl)methylpyrimidine homo-C-nucleoside,
using 2,4-dimethoxy-5-nitropyrimidine-6-ylmethylenetriphenyl-
phosphorane (<u>34</u>) as the condensing ylide.

The synthesis of compound <u>34</u> proved to be straight forward
and proceeded in two steps from 6-bromomethyl-2,4-dimethoxy-5-
nitropyrimidine[39] (<u>32</u>). Compound <u>32</u> was reacted with

triphenylphosphine to produce 2,4-dimethoxy-5-nitropyrimidin-6-ylmethyltriphenylphosphonium bromide (<u>33</u>) as a white solid which yellowed on standing. Without further purification <u>33</u> was treated with sodium hydroxide and converted to compound <u>34</u> in 79% yield from <u>32</u>. Compound <u>34</u> was very stable as evidenced by its high melting point (mp 205-206°C) and its long shelf life (6 months).

In order to investigate the feasibility of a condensation of <u>34</u> with the protected ribofuranose <u>16a</u>, a solution of the two reactants in a variety of solvents was heated at reflux. In acetonitrile only a minimal reaction was observed even after 72 hours. In benzene, a less polar solvent, only a slight improvement in the amount of detectable products was observed. However, in toluene a complete disappearance of starting sugar <u>16a</u> was observed after 72 hours. The use of <u>p</u>-xylene led to considerable darkening of the reaction mixture and decomposition of <u>16a</u>. When the progress of the condensation reaction in toluene was monitored on thin layer chromatography (tlc), the gradual disappearance of <u>34</u> and <u>16a</u> was accompanied by the appearance of 3 distinct products, 2,4-dimethoxy-6-<u>C</u>-(2,3-<u>O</u>-iso-propylidene-5-<u>O</u>-triphenylmethyl-1- α -D-ribofuranosyl)-methyl-5-nitroropyrimidine(<u>35</u>),2,4-dimethoxy-6-<u>C</u>-(2,3-<u>O</u>-isopropylidene-5-O-triphenylmethyl-1-ß-<u>D</u>-ribofuranosyl)methyl-5-nitropyrimidine (<u>36</u>) and a small amount of sodium methoxide to a solution of the product mixture in methanol, caused <u>37</u> to be rapidly converted into <u>35</u> and/or <u>36</u>. Interestingly, compound <u>36</u> was only sparingly soluble in methanol whereas <u>35</u> was quite soluble. This fact facilitated the isolation of pure <u>36</u> without the use of chromatography. Since both anomers were obtained in the reaction, comparison of [1]H-nmr data allowed us to make an unequivocal structure assignment. As expected the chemical

shift of the anomeric proton in 36 (4.37 ppm) occurred upfield
relative to that of 35 (4.84 ppm.).

The protected homo-C-nucleoside 36 was readily deblocked
using hydrochloric acid in dioxane to furnish what was assumed
to be 6-C-(1-ß-D-ribofuranosyl)methyl-5-nitropyrimidin-2,4-dione
(40) in 50% yield. In order to confirm the ß configuration,
the α anomer 35 was also deblocked using similar reaction
conditions. However, the product isolated was not the   anomer,
but rather the presumed ß anomer 40 in 87% yield. Finally the
ease with which the 5-nitro group in 35 is reduced was
demonstrated by the virtually quantitative, catalytic reduction
of 36 to afford 5-amino-2,4-dimethoxy-6-C-(2,3-O-isopropyli-
dene-5-O-triphenylmethyl-1-ß-D-ribofuranosyl)methylpyrimidine
(41).Unfortunately, the homo-C-nucleoside product 36 could not
be converted to compound 12 since various methods were
unsuccessful in introducing a formyl equivalent onto the
6-methylene carbon.

The second glycosylation procedure afforded the target
compound 7 in three steps starting from 6-cyanomethyl-2,4-
dimethoxy-5-nitropyrimidine (42). Compound 42 was selected as a
likely candidate for the condensation with 2,3-O-isopropylidene-
5-O-triphenylmethyl-D-ribofuranosyl chloride[40] (43) for three
reasons. First, it had the necessary nitrogen-containing moiety
residing at position 5 in the form of a nitro group. This group
was properly positioned to become the pyrrole nitrogen in the
final product. Second, the methylene protons at the six
position were sufficiently acidic, by virtue of the electron-
withdrawing potential of the cyano group as well as that of the
nitro-bearing pyrimidine ring, to allow malonate-like reactions
to occur. Diethyl malonate has been successfully condensed[41]
with compound 43 to yield the sugar diester derivative as a
ß-favoring anomeric mixture. The final reason for the

SCHEME 5

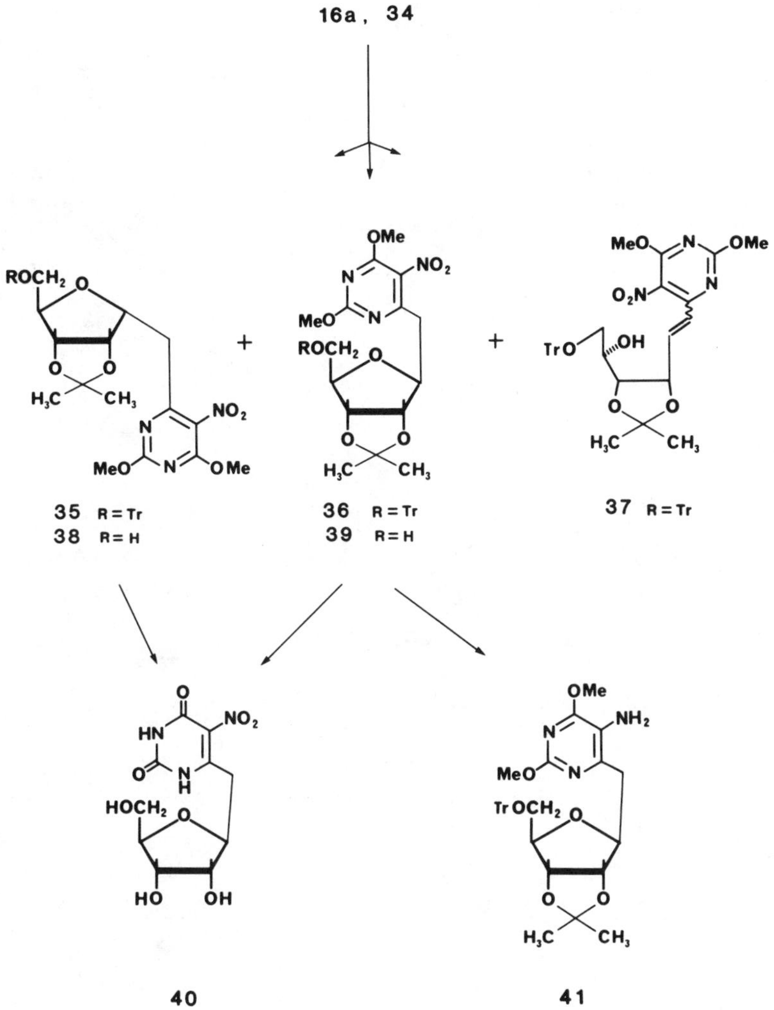

16a , 34

35 R = Tr
38 R = H

36 R = Tr
39 R = H

37 R = Tr

40

41

SCHEME 6

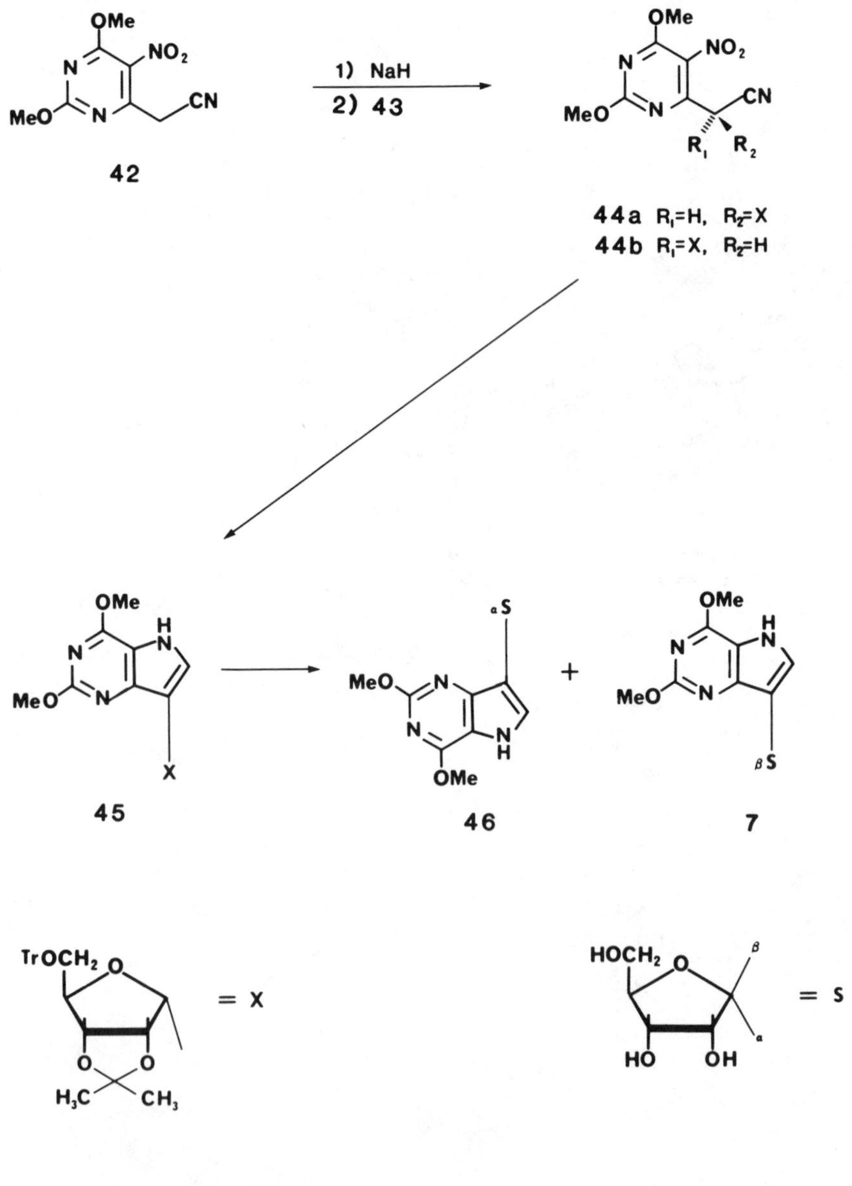
OMe
NO₂
N
MeO
N
CN
42
1) NaH
2) 43
OMe
NO₂
N
MeO
N
CN
R₁ R₂
44a R₁=H, R₂=X
44b R₁=X, R₂=H
OMe
H
N
N
MeO
N
X
45
αS
MeO
N
N
NH
OMe
46
+
OMe
H
N
N
MeO
N
βS
7
TrOCH₂
O
O
O
H₃C
CH₃
= X
HOCH₂
O
β
α
HO
OH
= S

selection of compound 42 involved the juxtaposition of the cyano and nitro substituents. It was felt that this arrangement would allow a catalytic reduction to form the pyrrolopyrimidine derivative in much the same way that o-nitro-cyanomethylbenzenes have been converted into indoles.[42,43]

Thus, compound 42, as a sodium salt, was reacted[44] with 43 in refluxing 1,2-dimethoxyethane in the presence of potassium iodide. When the reaction was interrupted after 90 minutes and worked up, a single, crystalline product, 2-(2,4-dimethoxy-5-nitropyrimidin-6-yl)-2-(2,3-O-isopropylidene-5-O-triphenyl-1-D-ribofuranosyl)acetonitrile (44) was obtained (Scheme 6). On the other hand, work-up of a similar reaction after a 4 hour reaction time resulted in the isolation of an inseparable mixture of the diasteromers 44a and 44b.

Compound 44a was catalytically reduced to 2,4-dimethoxy-7-C-(-2,3-O-isopropylidene-5-O-triphenylmethyl-1-D-ribofuranosyl)-pyrrolo[3,2-d]pyrimidine (45) by using 10% palladium on carbon. The removal of the sugar protecting groups in compound 45 was readily accomplished in methanolic hydrogen chloride and afforded 2,4-dimethoxy-7-C-(1-α -D-ribofuranosyl)pyrrolo[3,2-d]-pyrimidine (46) in 55% yield. Interestingly, when the deprotection was carried out in refluxing dioxane containing aqueous hydrochloric acid, a 1/1 mixture of 46 and the β anomer 7 was obtained. This apparent anomerization was accompanied by considerable decomposition resulting in a combined yield of only 12% for 46 and 7. The assignment of anomeric configuration was again made as a result of ^{1}H-nmr considerations.

In summary, two facile new approaches for the synthesis of pyrimidine homo-C-nucleosides have been presented. The first involved a Wittig condensation between the novel pyrimidyl-methylenephosphorane 34 and the protected ribofuranose 16a to afford compound 35. The second procedure involved the ribosylation of the active methylene carbon in compound 42 with the sugar $\underline{43}$ to produce the diastereomeric homo–C–nucleoside products $\underline{44a}$ and $\underline{44b}$. Although both methods led to the formation of either an anomeric mixture of products or predominantly the α anomer, subsequent chemical manipulations, which included a deblocking with aqueous acid, led to the formation of the ß anomer $\underline{7}$. We are currently investigating the possible uses of these procedures for the synthesis of additional C–nucleosides as well as new and novel homo–C– nucleosides.

REFERENCES

1. For recent reviews see: a) Hanessian, S.; Pernet, A. G.; Advances in Carbohydr. Chem. <u>1976</u>, <u>33</u>, 111; b) James, S. R. J. Carbohydr. Nucleotid. <u>1979</u> <u>6</u>, 417.

2. Hori, H.; Ito, E.; Takita, T.; Koyama, G. J. Antibiotics Ser. A <u>1964</u> <u>17</u>, 96.

3. Acton, E. M.; Ryan, K. J.; Henry, D. W.; Goodman, L. Chem. Commun. <u>1980</u>, 237.

4. Kalvoda, L. Coll. Czech. Chem. Commun. <u>1978</u>, <u>43</u>, 1431.

5. Buchanan, J. G.; Edgar, A. R.; Hutchison, J.; Stobie, A.; Wightman, R. H.Chem. Commun. <u>1980</u>, 237.

6. R. J. Suhadolnik, "Nucleoside Antibiotics,"Wiley– Interscience, New York,1970, chapter 9.

7. Sawa, T.; Fukagawa, Y.; Shimauchi, Y.; Ito, K.; hamada, M.; Takeuchi, T.; Umezawa, H. J. Antibiot. Ser. A <u>1965</u>, <u>18</u>, 259.

8. Sawa, T.; Fukagawa, Y.; Shimauchi, Y.; Ito, K.; Hamada, M.; Takeuchi, T.; Umezawa, H. J. Antibiot. Ser. A <u>1967</u>, <u>20</u>, 317.

9. Sheen, M. R.; Kim, B. K.; Martin, H.; Parks Jr., R. E.
    Proc. Am. Assoc. Cancer Res. 1968, 9, 249. See also:
    Tsukuda, I.; Kurimoto, T.; Hori, Komai, T. J. Antibiot.
    Ser. A 1969, 22, 36.

10. Crabtree, G. W.; Agarwal, R. P.; Parks Jr., R. E.; Lewis,
    A. F.; Wotring, L. L.; Townsend, L. B. Biochem. Pharmacol.
    1979, 28, 1491.

11. Lim, M.-I.; Klein, R. S. Tetrahedron Lett. 1981, 22, 25.

12. Lim, M.-I.; Klein, R. S.; Fox, J. J. Tetrahedron Lett.
    1980, 21, 1013.

13. Chu, M. Y.; Landry, L. B.; Klein, R. S.; Lim, M.-I.;
    Bodgden, A. E.; Crabtree, G. W. Proc. Am. Assoc. Cancer
    Res. 1982, 23, 220.

14. Gupta, C. M.; Hope, A. P.; Jones, G. H.; Moffatt, J. G.
    175th ACS National Meeting, Anaheim, CA, March 13-17, 1976,
    Abstr. CARB. 40.

15. Recondo, E. F.; Rinderknecht, H. Helv. Chim Acta 1959,
    42, 1171.

16. Secrist III, J. A. J. Org. Chem. 1978 43, 2925.

17. Watanabe, K. A.; Hollenberg, D. H.; Fox, J. J. J.
    Carbohydr. Nucleotid. 1974, 1, 1.

18. Ogawa, T.; Pernet, A. S.; Hanessian, S. Tetrahedron Lett.
    1973, 3543.

19. a) D. J. Brown, "The Pyrimidines," Wiley, New York, N.Y.,
    1962, Chapter II, p. 48; b) D. J. Brown, "The Pyrimidines.
    Supplement I," Wiley, New York, N. Y., 1970, Chapter II, p.
    31.

20. a) D. J. Brown, "The Pyrimidines," Wiley, New York, N. Y.,
    1962, Chapter II, p. 65; b) D. J. Brown, "The Pyrimidines.
    Supplement I," Wiley, New York, N. Y., 1970, Chapter II, p.
    47; c) Ito, I.; Ora, N.; Kato, T.; Ota, K. Chem. Pharm.
    Bull. 1975, 23, 2104.

21. For a review see: Amarnath, V.; Madhav, R. Synthesis 1974, 837.

22. For example see: Garg, H. G.; Joshi, S. S. J. Indian Chem. Soc. 1960, 37, 626.

23. Klein, R. S.; Lim M.-T.; Tam, S. Y-K; Fox, J. J. J. Org. Chem. 1978, 43, 2536.

24. Ogawa, T.; Pernet, A. S.; Hanessian, S. Tetrahedron Lett. 1973, 3543.

25. Systematically this compound should be referred to as 7,10-anhydro-1,2,3,4,5,6-hexadeoxy-8,9,11-tri-O-benzoyl-D-allo-(altro)-undec-4-ene but, for ease of discussion, it and other structurally related compounds will be referred to by more immediately obvious trivial names.

26. The term anomeric is not strictly applicable to C-glycosides but is used for convenience.

27. Noyori, R.; Sato, T.; Hayakawa, Y. J. Am. Chem. Soc. 1978, 100, 2561.

28. DeBernardo, S.; Sato, T.; Hayakawa, Y. J. Am. Chem. Soc. 1978, 100, 2561.

29. Lim, M.-I.; Klein, R. S. Tetrahedron Lett. 1981, 25.

30. Lim, M.-I.; Klein, R. S.; Fox, J. J. Tetrahedron Lett. 1980, 1013.

31. Chu, C. K.; Wempen, I.; Watanabe, K. A.; Fox, J. J. J. Org. Chem. 1976, 41, 2793.

32. Cupps, T. L.; Wise, D. S.; Townsend, L. B. Carbohydr. Research 1982, Accepted for publication.

33. Hosomi, A.; Endo, M.; Sakurai, H. Chem. Lett. 1976, 941; Ojima, I.; Kumagai, M. Chem. Lett. 1978, 575.

34. Tsunoda, T.; Suzuki, M.; Noyori, R. Tetrahedron Lett. 1980, 71.

35. Cupps, T. L.; Wise, D. S.; Townsend, L. B. J. Org. Chem. 1982, 47, 5115.

36. Krauss, J. C.; Cupps, T. L.; Wise, D. S.; Townsend, L. B.
     Synthesis 1982, Accepted for publication.

37. (a) Ohrui, H.; Jones, G. H.; Moffatt, J. G.; Maddox, M. L.;
     Christensen, A. T.; Bryam, S. K. J. Am. Chem. Soc. 1975,
     97, 4602; (b) Cousineau, T. J.; Secrist III, J. A. J.
     Carbohydr. Nucleosid. Nucleotid. 1976, 3, 185.

38. (a) Hanessian, S.; Ogawa, T.; Guindon, Y. Carbohydr.
     Research 1974, 38, C12.

39. Cupps, T. L.; Wise, D. S.; Townsend, L. B. Tetrahedron
     Lett. 1982, 23, 4759.

40. Klein, R. S.; Ohrui, H.; Fox, J. J. J. Carbohydr.
     Nucleosid. Nucleotid. 1974, 1, 265.

41. Ohhrui, H.; Jones, G. H.; Moffatt, J. G.; Maddox, M. L.;
     Christensen, A. T.; Byran, S. K. J. Am. Chem. Soc. 1975,
     97, 4602.

42. Walker, G. H. J. Am. Chem. Soc. 1955, 77, 3844.

43. Cupps, T. L.; Wise, D. S.; Townsend, L. B. J. Org. Chem
     1982, Accepted for publication.

44. Cupps, T. L.; Wise, D. S.; Townsend, L. B. J. Org. Chem.
     1983, Accepted for publication.

Index